T0133289

GUIDE TO COMETS

# CONTENTS

## LIST OF PLATES

ACKNOWLEDGEMENTS

My grateful thanks are due to Commander H. R. Hatfield, E. A. Whitaker, Jack Bennett, D. G. Daniels, Jack McBain, K. Kennedy and the late Dr. E. M. Lindsay for allowing me to see their photographs. I also thank the Mount Wilson and Palomar Observatories and the Royal Greenwich Observatory, Herstmonceux. Lawrence Clarke has undertaken the task of producing the line drawings; and last but by no means least I must thank Michael Foxell, of Lutterworth Press, for his invaluable help and encouragement.

P.M.

Selsey. *May 1977*

*Chapter One*

# VISITORS FROM SPACE

When beggars die, there are no comets seen:
The heavens themselves blaze forth the death of princes.

(*Julius Caesar*, II 2)

THE WORDS ARE SHAKESPEARE'S; the belief is very old indeed. For many centuries, and indeed until relatively modern times, comets were regarded as messengers of disaster. There were even suggestions that a collision with a comet could bring about the end of the world. Certainly the spectacle of a great comet, with a brilliant head and a tail stretching half-way across the sky, could well be expected to strike terror into the hearts of primitive peoples; but the fear of comets lingered on for a surprisingly long time, and is still not entirely dead. On my desk I have a pamphlet headed *The Christmas Monster*—a reference to a comet (Kohoutek's) which was expected to become really bright toward the end of 1973. The pamphlet, put out by a religious sect, made all manner of dire forecasts, including world-wide destruction. It was widely distributed, and no doubt there were some people who took it seriously, even though Kohoutek's Comet finally turned out to be a great disappointment as a spectacle.

Actually, there is no reason whatsoever for anyone to be alarmed by comets. They are harmless, and they are fascinating; by now we have learned a great deal about them, and plans are being made to send up a space-probe to rendezvous with a suitable comet. Any such idea would have seemed fantastic only a few decades ago, but there is nothing far-fetched about it today.

What I hope to do, in this book, is to give a general account of modern cometary astronomy. First, then, what exactly is a comet—and do comets belong to the Sun's family, or Solar System?

The Solar System is made up of one star (the Sun), nine principal planets (of which the Earth comes third in order of distance), and various bodies of lesser importance, such as the satellites which attend some of the planets. We have one natural satellite, our familiar Moon, which keeps company with us in our never-ending journey round the Sun. It is less than a quarter of a million miles from us, so that astronomically speaking it is on our doorstep. Recently I checked the dashboard of my elderly Ford Prefect car, and found that the mileage covered during the past twenty-five years exceeded 600,000. This means that I have driven more than twice the distance to the Moon.

Yet the Solar System is built upon a grand scale. The Sun, at 93,000,000 miles from us, is nothing more than an average star—much less luminous than many of the stars visible on any clear night—but it is the ruler of our own particular part of the universe, and it keeps the planets firmly under control. The planets themselves move in paths or orbits which are not very different from circles, and they shine by reflecting the light of the Sun, so that they appear starlike. Venus, Jupiter and Mars may outshine even the most brilliant of the stars, while Saturn also is a conspicuous object, and Mercury is not hard to see in the dusk or dawn sky when it is best placed. The remaining planets (Uranus, Neptune and Pluto) are fainter, and were discovered in what we may call modern times: Uranus in 1781, Neptune in 1846 and Pluto as recently as 1930. Pluto, incidentally, has an orbit which is relatively eccentric, and it may not be worthy to rank as a true planet at all. There are suggestions that it is merely an ex-satellite of Neptune, which managed to break free and move off independently.

The night-time stars are much more remote than the Sun. Even the nearest of them is more than 24 million million miles away, so that its light, moving at 186,000 miles per second, takes 4·2 years to reach us; we say, therefore, that the distance of this particular star (Proxima Centauri) is 4·2 light-years. Most of the rest are more remote still; for example, Rigel in Orion lies at something like 900 light-years, so that we see it today as it used to be in the time of William the Conqueror. This explains why the stars look so much fainter than the Sun, and why their individual movements seem very slow. The star-patterns or constellations appear virtually the same today as they must have done to the builders of the Egyptian Pyramids. The planets, which are much closer to us, wander about from one constellation to another, though admittedly they keep within certain well-defined limits. The word 'planet' comes, indeed, from a Greek word meaning 'wanderer'.

The Solar System is divided into two main parts. There are four comparatively small, solid planets (Mercury, Venus, the Earth and Mars) with distances from the Sun ranging between 36 million miles and 141 million miles. (I am here giving mean values, suitably rounded off.) Then comes a wide gap, in which move thousands of dwarf worlds known as the minor planets or asteroids; even the largest of them, Ceres, is much smaller than our Moon, and is no more than 700 miles in diameter at most. Beyond the asteroids come the four giants: Jupiter, Saturn, Uranus and Neptune, with distances ranging from 483 million miles for Jupiter out to 2793 million miles for Neptune. Pluto, which is certainly smaller than the Earth, appears to be in a class of its own. At its nearest point to the Sun it comes closer-in than Neptune, but its orbit is tilted at the relatively sharp angle of 17 degrees, so that there is no fear of a collision on the line.

It would be wrong to suppose that these are the only bodies in the Solar System. There are many more. To begin with, we have the satellites or moons; Jupiter has fourteen, Saturn ten, Uranus five, and Neptune and Mars two each, while the Earth has a solitary attendant. Also there are countless tiny particles, known collectively as meteoroids, which we cannot see unless they come so close to the Earth that they dash into the upper atmosphere. Finally, there are those strange, erratic travellers—the comets.

I think it may be best to clear up one point immediately. I have had many letters from people who say, in various ways: 'Last night I saw a bright object flashing across the sky. Can it have been a comet?' The answer is, quite definitely: 'No'. A comet lies well beyond the limits of the Earth's atmosphere, and is in orbit round the Sun. This means that it does not move perceptibly over a few seconds or even a few minutes, and one has to watch it for hours before detecting any motion against the starry background. So if you see an object moving quickly along, it cannot be a comet. If astronomical, it must be either a meteor or else an artificial satellite.

In 1957 the Russians launched the first orbital vehicle in history, Sputnik 1; since then many hundreds have been sent up, and because they too reflect the sunlight they look like moving stars. A few, such as the balloon Echo satellites of the 1960s, can be very brilliant, and have given rise to scores of flying saucer reports. Artificial satellites are outside the scope of a book about comets, but the important thing to remember about these satellites is that if they are put into stable orbits round the Earth, far enough out to avoid friction against the upper air, they move in precisely the same way as natural astronomical bodies would do. This also applies to the true space-stations, of which Skylab of 1973 was the first.

Meteors, however, are relevant because they are closely associated with comets—as we will see later. A meteor is a tiny particle, of sand-grain size or less, moving round the Sun. If it comes close enough to the Earth (below 120 miles or so) it has to force its way through the air-particles. This sets up friction, generating so much heat that the meteor destroys itself, ending its career in the streak of light which we call a shooting-star. When the Earth passes through a swarm of meteors, as happens several times in each year, we see a shower of shooting-stars. August is a particularly good 'meteor month', and anyone who stares upward at a dark, cloudless sky for a few minutes at any time during the first fortnight of August will be very unlucky not to see a meteor or two.

Now and then we encounter larger objects, which can survive the complete drop to ground level without being destroyed, and are then known as meteorites. Most museums have meteorite collections—and so have some planetaria, notably the Hayden Planetarium in New York, where you can see a huge meteorite weighing over thirty tons. This particular specimen was found in Greenland by the American polar explorer Robert Peary, and remains the largest meteorite 'in captivity'. The largest known meteorite is still lying where it fell, at Hoba West, near Grootfontein in Southern Africa; I doubt whether anyone will try to run away with it, since the total weight exceeds sixty tons. Yet it seems clear that meteorites are basically different in nature from meteors, and are more closely related to the asteroids.

Comets, of course, are different again, and let it be stressed at once that they are not nearly so important as they sometimes look. There have been comets brilliant enough to be seen in broad daylight, but their mass is very small compared with that of a planet, or even a satellite such as the Moon. It has been said that a comet is 'the nearest approach to nothing that can still be anything'; we are dealing with a body made up of small solid particles, ices of various kinds, and extremely thin gas.

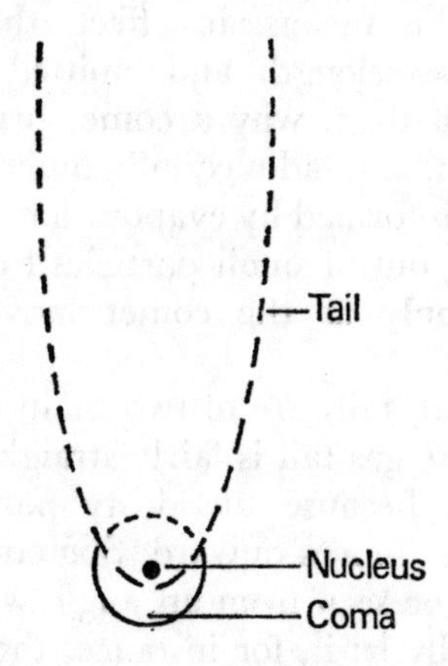


Fig. 1. Physical structure of a comet, showing the nucleus, coma, and tail. Not all comets have visible nuclei, and many of them have no tails

A bright comet is composed of three main parts (Fig. 1). First there is the nucleus, which contains most of the total mass, but cannot be more than a few miles in diameter (and it is by no means certain that all comets have well-defined nuclei). Around this is the head, or coma, which contains small particles made up of materials such as iron, nickel and magnesium, together with icy substances in which frozen water, ammonia and methane are thought to be the most important constituents. In some cases the coma may be huge; that of the Great Comet of 1811 had a diameter of a million and a quarter miles, so that it was much bigger than the Sun. Extending away from the coma there may be a tail, composed of even smaller particles ('dust') and of incredibly tenuous gas. Some comets have tails of both kinds—that is to say, 'gaseous' and 'dusty'. The Great Comet of 1843 had a main tail which was 200 million miles long, which is greater than the distance between the Sun and Mars. On the other hand, many small comets have no tails at all, and look like nothing more than small, fuzzy patches in the sky. It has been said that they give the impression of dimly-shining cotton-wool.

Left to itself, a comet would not shine at all, but when it nears the Sun it is lit up, and we see it by reflection. In the thin stuff of the coma we also have the phenomenon known as fluorescence. This means, in effect, that the sunlight is absorbed at one wavelength and emitted at another. It is easy to understand, then, why a comet brightens up quickly as it comes sunward, and fades equally quickly when it recedes. Moreover, the tail is formed by evaporation of the icy materials and by the driving-out of small particles from the nucleus, so that it develops only as the comet moves into the Sun's neighbourhood.

As we have noted, tails are of two main types, gaseous and dusty. Generally the gas-tail is fairly straight and the dust-tail noticeably curved, because the dusty particles tend to lag behind as the comet moves onward. Sometimes, of course, the tail (or tails) may be seen from an angle which makes it look shorter than it really is; if, for instance, the comet is heading straight toward us, the tail will not be seen to best advantage. Tail structure can sometimes be very complex, and there are rapid, involved changes, with distinct condensations, jets and streamers. Despite its flimsy nature, a comet is anything but inactive.

One very interesting fact is that a tail always points more or less away from the Sun, as shown in Figure 2, so that during an outward journey the comet actually travels tail-first. This has been known for a long time, and various theories have been put forward to account for it. One suggestion, proposed some seventy years ago, involved the pressure of sunlight. Light does exert a pressure, and a light-source will tend to push objects away from it, as was proved experimentally in 1900 by the Russian physicist Pyotr Lebedev. By everyday standards the force is very feeble; the repelling force of sun-

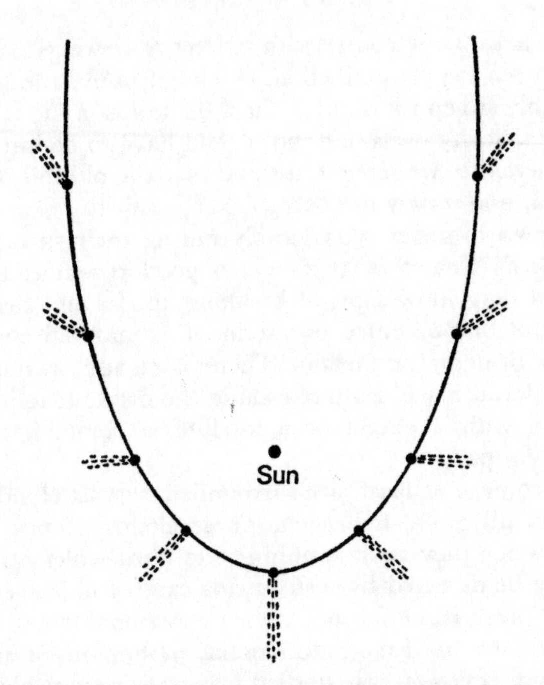


Fig. 2. A comet's tail always points more or less away from the Sun because of the repelling effect of solar wind, i.e., the low-energy particles being sent out from the Sun constantly in all directions. When travelling outward, therefore, the comet moves more or less tail first

light on a square foot of the Earth's surface can never be more than one ten-millionth of a pound, which is not very much! But with a comet's tail, we are dealing with very tiny particles, and we have to balance radiation pressure against the attractive force of gravitation. It was calculated that for particles about 1/100,000 of an inch in diameter, radiation pressure would win the tug-of-war, so that the 'dust' in a comet's tail would be driven outward from the nucleus.

Recently there has been a change in attitude, and theorists have tended to look toward what is called 'solar wind'—a constant stream of electrified, low-energy particles streaming out from the Sun in all directions. Most astronomers now believe that the solar wind exercises the main influence upon the direction of a comet's tail, although the full details about what is happening are still rather uncertain.

When a brilliant comet with a long tail swings round the Sun as it reaches its perihelion, or closest point, the tail must presumably sweep round too; but if the tail is millions of miles long, the velocity of its 'far end' would have to be improbably high. Therefore, we have to assume that the old tail is largely destroyed, and a new one formed fairly quickly. Comets have been known to suffer considerably during their rush past the Sun—West's Comet of 1976 was a good recent example of this—and may show signs of breaking up. In any case, there is no doubt that a comet loses some of its material every time it passes through perihelion. There is steady wastage, and dusty material is left scattered along the orbit. It follows that compared with a planet or a satellite, a comet has a very limited life-span.

Some comets, at least, are surrounded by vast clouds of the lightest of all gases, hydrogen. These clouds cannot be seen directly, since they radiate only in the ultra-violet range, but they may be detected by instruments carried in space-probes. The first proof came in 1968, when a reasonably conspicuous comet, known as Tago-Sato-Kosaka in honour of its three Japanese discoverers, was studied from the space-vehicle OAO2 (Orbiting Astronomical Observatory No. 2) and was found to have a hydrogen cloud a million miles in diameter. Similar clouds were found with Bennett's Comet of 1970 and Kohoutek's Comet of 1973, which was, incidentally, the first comet to be studied from a manned satellite (Skylab). There is no reason to doubt that clouds of the same kind are associated with other comets. All in all, a comet is a much more complicated structure than it might seem.

Obviously, the telescope is the astronomer's essential research tool, but he would be lost without the spectroscope, which splits up light and tells what substances exist in the light-source itself. Spectroscopy can show which elements are present in the Sun, the other stars, the great gas-clouds or nebulæ, and even the galaxies which are so far away that their light takes millions of years to reach us. We have come a long way since 1830, when the French philosopher August Comte predicted that the chemistry of the stars would forever remain a mystery to mankind.

The first attempt to study the spectrum of a comet was made

in 1864 by an Italian astronomer, Giovanni Battista Donati. When he turned his spectroscope toward a comet which happened to be on view, he made a fascinating discovery. Donati expected to see a weak solar spectrum—and he did, but he also saw effects which could be due only to light emitted by the cometary gases on their own account. Donati could take matters little further, since in those days spectroscopy was in its infancy, but by now we have been able to identify many different substances. Some of the gases would be toxic if dense, but the gas in a comet is many millions of times more rarefied than the air that you and I are breathing, so that it can do no damage at all. Even a direct collision with the nucleus of a comet would produce only local effects—and, as we will see later, there is excellent evidence that a small comet really did hit the Earth in 1908.

Various theories about the make-up of comets have been put forward. One of these, due mainly to the British astronomer R. A. Lyttleton, is that of the 'flying gravel-bank'. According to Lyttleton, both nucleus and coma are made up of dust-particles, concentrated toward the centre of the comet and making the nucleus appear deceptively solid. As the comet nears the Sun, and is warmed, the gases inside the particles escape, and a tail is formed. Each particle is moving round the Sun in its own independent path, and near perihelion the whole structure becomes more crowded; particles collide with one another and are pulverized, producing even more finely divided material which is driven out by the pressure of the Sun's radiation and by the solar wind.

Recent research has cast serious doubt upon the gravel-bank theory, and most astronomers have more faith in the 'dirty ice-ball' idea, developed by F. L. Whipple in the United States. Rather than being made up of a swarm of individual particles, the nucleus—which, of course, contains most of the mass of the comet—is composed of what is called a conglomerate. The word comes from the Latin *conglomerare*, to form into a ball; it is used by geologists to describe pieces of fragmented rock held together in finer-grained material. On the Whipple model, a comet's nucleus is composed of a conglomerate of rocky fragments held together with 'ices' —frozen methane, ammonia, carbon dioxide and other substances, including water. When the comet approaches perihelion, the ices begin to evaporate. Methane is the first to do so; the others follow, and this at least explains the rapid, irregular changes, notably the pronounced jets which sometimes issue from the nucleus. In 1974 it was found that Kohoutek's Comet contained large quantities of the ionized water molecule known to chemists as $H_2O^+$. This identification of water in a comet clearly supports the dirty ice-ball rather than the flying gravel-bank theory.

Whichever picture is right—or even if both are wrong—comets are unlike any other bodies in the Solar System. We do not yet understand them fully, but the prospects are bright. When we manage to send a space-probe through one of these strange, eerie visitors from afar, many of their outstanding problems should be solved.

*Chapter Two*

# THE LORE OF COMETS

BECAUSE COMETS MAY BECOME so spectacular, records of them go back for many centuries. The same is true of total eclipses of the Sun, when for a few minutes 'day is turned into night' as the Moon covers up the brilliant solar disk. Before true science began, events of this sort caused considerable alarm, which was hardly surprising.

To the ancient Chinese, eclipses were due to a hungry dragon which was trying to gobble up the Sun, and had to be scared away by the banging of gongs and drums and by making as much noise as possible—a procedure which, let it be added, always worked! But an eclipse is a brief affair, whereas a comet can loom balefully in the sky for several days, weeks, or occasionally months. This is something that no younger people of today can appreciate to the full, because it is so long since we had a comet of this kind. One or two seen in recent years have become brilliant (the comet of 1965 particularly so), but have not remained in view for long. Things were different in the nineteenth century, when there were several really imposing visitors—in 1811, 1843, 1858, 1861 and 1882, among others.

Comets were originally regarded as signs of divine displeasure, and there were no obvious ways of pacifying the gods; no amount of gong-beating would make a comet 'go away'. Even the Greeks had no idea of their true nature. Anaxagoras, who was born about 500 B.C., had remarkable insight, and believed the Sun to be a red-hot stone larger than the Peloponnesus, the peninsula upon which Athens stands; but he thought that comets were produced by the clustering of faint stars. Around 350 B.C. the famous Aristotle taught that both comets and meteors were due to 'hot, dry exhalations' rising from the ground and being carried along by the motion of the sky, so becoming heated and bursting into flame; slow movement produced a comet, rapid movement resulted in a meteor. Aristotle's authority was so great that for many centuries after his death the idea of comets as upper-air phenomena was generally accepted.

There was also the astrological aspect. In Roman times the great writer Pliny commented that 'We have, in the war between Cæsar and Pompey, an example of the terrible effects which follow the apparition of a comet. Toward the commencement of this war, the darkest nights were made light, according to Lucan, by unknown stars; the heavens appeared on fire; burning torches traversed in all directions the depths of space; the comet, that fearful star, which overthrows the powers of the Earth, showed its terrible locks.' Later, Vespasian —one of the better emperors—refused to be alarmed by the comet of A.D. 79, and is said to have remarked: 'This hairy star does not concern me; it menaces rather the King of the Parthians, for he is hairy, while I am bald.' Alas, Vespasian's confidence turned out to be misplaced, and he died in the same year. Another emperor to make his exit during the visibility of a bright comet was Macrinus, in A.D. 218. Macrinus, however, was no Vespasian; he was merely one of the succession of nonentities who occupied the throne during the period following the death of the philosopher-emperor Marcus Aurelius. Incidentally, the comet of 218 was Halley's, about which I will have much more to say in Chapter 6. (Halley's Comet was also on view in 1066, when the Normans were preparing to invade England, and in 1456, when Pope Calixtus III actually excommunicated it.)

The first real advance in cometary studies was due to the great Danish astronomer Tycho Brahe, in 1577. Tycho was an extraordinary man, but he was much the most skilful observer of pre-telescopic times, and for twenty years he worked away on his island observatory at Hven, in the Baltic, drawing up an accurate star-catalogue and measuring the apparent movements of the planets. It was his work which enabled his last assistant, Kepler, to prove that the Earth goes round the Sun instead of vice versa—something which Tycho could never bring himself to believe. While at Hven, Tycho observed seven comets. That of 1577 was fairly bright, and Tycho set out to discover whether it showed any measurable 'diurnal parallax'. If it were closer than the Moon (whose distance was known with fair accuracy), it ought to show a shift against the starry background as the Earth turned on its axis; but it did not, and Tycho calculated that the comet must be at least six times as far away as the Moon. As we now know, it was considerably more remote than that, but there was nothing the matter with Tycho's reasoning.

The sequel was rather strange. In 1618 an Italian named Orazio Grassi published a book in which he repeated Tycho's conclusions, and he was bitterly attacked by no less a person than Galileo, who was the first great observer to use a telescope, and who played a major rôle in destroying the old theory that the Sun goes round the Earth. The story of how Galileo was subsequently brought up before the Inquisition and forced into a public and absolutely hollow recantation is so well known that I need not repeat it here; but even though he was right about the motion of the Earth he was quite wrong about comets, which he thought to be due to the refraction of sunlight in vapours rising from the Earth. In this case, of course, a comet would not show parallax, any more than a rainbow does. It is curious that a man of his perception should have fallen into such a trap, but he never changed his opinion.

Despite Galileo, the idea of comets as true celestial objects was gaining ground, even though their nature was still obscure. Descartes, the French philosopher who worked out a 'theory of vortices' for the universe, supposed comets to be stars which gradually became covered with spots until they lost their light, after which they could not keep their set places in the sky, and were carried along until they approached the Sun closely enough to be lit up and made visible to us. The real change in outlook came with Edmond Halley, who calculated that one particular comet appeared regularly every seventy-six years, and was therefore a bona-fide member of the Solar System, travelling in an elliptical path which brought it back time and again to the neighbourhood of the Sun. The first predicted return of Halley's Comet, in 1758–59, ushered in what we may call the modern era.

The fear of comets was both superstitious and practical. That rather shadowy and elusive Englishman, Thomas Digges, wrote in 1556 that 'Comets signify corruption of the stars. They are signs of earthquakes, of wars, of changing of kingdoms, great dearth of corn, yea a common death of man and beast.' (I have modernized his spelling.) There was also the remarkable proclamation issued by the town council at Baden, in Switzerland, when a comet with 'a frightful long tail' appeared in 1681. The town authorities ordered that 'all are to attend Mass and Sermon every Sunday and Feast Day, not leaving the church before the sermon or staying away without good reason; all must abstain from playing or dancing, whether at weddings or on other occasions; none must wear unseemly clothing, nor swear nor curse'. Whether these orders were obeyed to the letter we do not know. If so, that particular period must have been rather dull for the Badenese. So far as I have been able to discover, the last pronouncement of this kind by a serious astronomer was made on 4 July 1816 by a certain Dr. Pennada, addressing the Padua Institute in Italy. Pennada claimed that 'the most remarkable political events have always been preceded, accompanied or followed by extraordinary astro-meteorological phenomena'—a clear reference to comets.

It would be wrong to say that comets have exerted a great influence upon human thought. They have not; they were merely symbolic of the general feeling that Man's destiny was ruled, and even forecast, by events in the sky. But quite apart from these intangible fears there was also alarm at the possible effects of a cometary collision, and there have been occasional panics on this score.

Probably the most famous of such predictions was made by the scholar and clergyman Dr. William Whiston, who was contemporary with Sir Isaac Newton, and succeeded Newton as Lucasian Professor of Mathematics at the University of Cambridge, in 1710. Fourteen years earlier he had published his book *A New Theory of the Earth*, in which he tried to link the Book of Genesis with the wanderings of a comet which periodically approached the Sun and was violently heated. As the orbit became more circular, said Whiston, the overheating stopped, and the comet became a planet. Man appeared, but the grievous sins of humanity caused another comet to strike the Earth, giving it a movement of rotation and producing a huge tide which was, of course, the Biblical Flood. Whiston gave the date of this disaster as either 28 November 2349 B.C. or 2 December 2926 B.C. He went on to claim that the comet

# AN ALLARM TO EUROPE:

By a Late Prodigious

# COMET

ſeen *November* and *December*, 1680.

With a Predictive Diſcourſe. Together with ſome preceding and ſome ſucceeding Cauſes of its ſad Effects to the *Eaſt* and *North Eaſtern* parts of the World.

Namely, *ENGLAND*, *SCOTLAND*, *IRELAND*, *FRANCE*, *SPAIN*, *HOLLAND*, *GERMANY*, *ITALY*, and many other places.

---

By *John Hill* Phyſitian and Aſtrologer.

---

The Form of the *COMET* with its Blaze or Stream as it was ſeen *December* the 24th. Anno 1680. In the Evening.

London Printed by *H. Brugis* for *William Thackery* at the Angel in Duck-Lane.

Fig. 3. Title page of *An Allarm to Europe*, written in 1680 by John Hill

would eventually return, this time changing the Earth's path again and transforming the world once more into a comet—with the inevitable destruction of all life.

Whiston's subsequent career was decidedly chequered, and he eventually lost his post as Lucasian Professor; but his scientific reputation made at least some people take his theories seriously. In fact, of course, they were quite absurd. The difference in mass between a comet and even a small planet is tremendous, quite apart from their differences in nature, so that there is not the slightest chance of a comet changing into a planet or vice versa. And yet even in our own time there has been a revival of the idea. Since the 1950s books have flowed from the pen of a psychiatrist, Dr. Immanuel Velikovsky, who was born in Russia, trained in Israel and has spent much of his life in America. Velikovsky's first offering, entitled *Worlds in Collision*, was in some respects similar to Whiston's inasmuch as it confused planets with comets. It was widely publicized, and started a curious cult which still persists.

To Velikovsky, the time-scale of the universe is measured in thousands of years rather than in millions. He explains how the giant planet Jupiter suffered a tremendous outburst, and shot out a comet which later became the planet Venus. In 2500 B.C. the comet Venus bypassed the Earth; this was the time of the Israelite Exodus led by Moses, and the Earth's rotation was slowed down, drying up the Red Sea at a convenient moment for the Israelites to cross. Tremendous upheavals followed; petrol rained down, so that our modern fuel represents 'remnants of the intruding star which poured forth fire and sticky vapour'. Two months later, after the Earth had started spinning again, the comet Venus returned for a second visit, producing the thunder and lightning noted at the time of the sermon on Mount Sinai. Other encounters followed, one of which caused the tremors which shook down the walls of Jericho, but then the comet collided with Mars and lost its tail, so turning into the modern planet Venus. Meanwhile Mars itself moved closer to the Earth, and almost hit us in 687 B.C. . . . Velikovsky gives instance after instance, all 'supported' by quotations from the Bible which are, incidentally, quite correct!

Nobody with even the most elementary knowledge of science can take this rigmarole seriously, but at least Velikovsky's theories are harmless, and the world would be poorer without the never-ending supply of what I have termed 'Independent Thinkers'.*

Going back in time, I must refer briefly to the French panic of 1773, which arose from the misinterpretation of a scientific paper written by the eminent mathematician Joseph Jerome de Lalande. People jumped to the conclusion that a comet would collide with the Earth on 20 or 21 May, and there were numerous sales of seats in Paradise, which the clergy were said to have obtained by special dispensation. There was yet another alarm in 1832, this time connected with Biela's periodical comet. Calculations showed that the comet would cross the Earth's orbit on 29 October, but as the Earth itself was nowhere near that point in its path there was no chance of even a close encounter. The distance between the Earth and the comet was always well over forty million miles.

End-of-the-world scares still occur in backward countries. On 30 June 1973 there was a total eclipse of the Sun visible from parts of Africa, and a team of European scientists arriving in Kenya were given an unfriendly reception by the local inhabitants, who believed that the White Men were trying to kill the Sun. A comet may not be so startling as a total eclipse, but it certainly lasts for longer, and no doubt the witch-doctors will have their say next time a brilliant comet appears. Bennett's Comet of 1970 caused considerable alarm in Arab countries, since it was suspected of being an Israeli war weapon.

It is also worth recalling that in 1910, when Halley's Comet was last on view, an enterprising manufacturer in America made a large sum of money by selling what he called comet pills, though I have no idea what they were meant to do. Human nature is slow to change.

* I have discussed Velikovsky's theories, and others of their kind, in my book *Can You Speak Venusian?* (Star Books; paperback, 1976).

*Chapter Three*

# THE PATHS OF COMETS

AS WE HAVE NOTED, comets used to be regarded as luminous effects in the Earth's upper air. This belief lasted until 1577, when Tycho Brahe showed that the comet of that year lay well beyond the Moon. But it was not until the time of Newton and Halley, in the late seventeenth century, that anyone had any real idea of how a comet must move; at that time, of course, it had only just been realized that the Sun, rather than the Earth, is the central body of the Solar System.

According to ancient theories, all celestial bodies were assumed to be circular, because the circle is the 'perfect' form, and nothing short of absolute perfection has any right to be in the heavens. The man who overthrew this concept was Johannes Kepler, whose famous Laws of Planetary Motion were published between 1609 and 1618. Because the Laws are so important, it seems worth while to describe them here in some detail.

The first Law states that a planet moves round the Sun in an ellipse, not in a circle. A typical ellipse of low eccentricity is shown in the first diagram (Fig. 4). The two points lettered F and F¹ are the 'foci' of the ellipse, and are reasonably close together, so that the ellipse is not very markedly different from

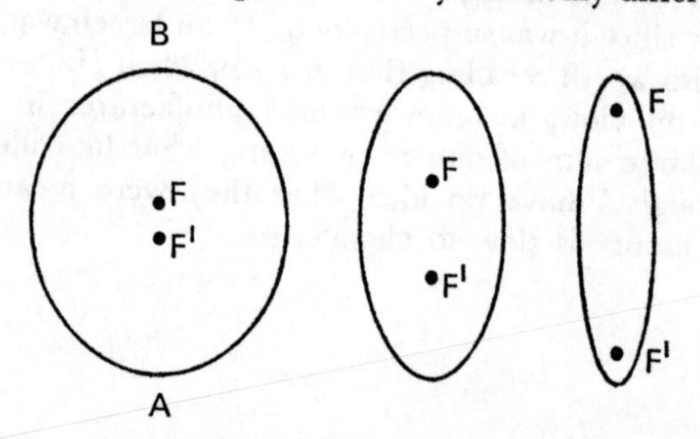


Fig. 4. (a) An ellipse of low eccentricity. F and F¹ indicate the two foci. (b) An ellipse of higher eccentricity, such as the orbit of a typical short-period comet. (c) An ellipse of high eccentricity, such as the orbit of a comet of a longer period

a circle. With a planetary orbit, the Sun lies in one focus of the ellipse, while the other focus is empty.

The ratio of the distance F–F[1] to the whole length of the ellipse (AB) is a measure of the eccentricity. If F and F[1] coincide, the eccentricity is zero, and the result is a circle. In other words, a circle is simply an ellipse with no eccentricity at all. If the distance between the foci (F and F[1]) is one-third of the distance between A and B, the eccentricity will be 0·33, as shown in the second diagram; and if the eccentricity is still greater—say 0·8—the ellipse will be long and narrow (third diagram).

The orbits of all the principal planets are of low eccentricity. For instance, the Earth's distance from the Sun ranges between 91½ million miles at closest approach (perihelion) to 94½ million miles at greatest recession (aphelion). Not so with the comets, most of which have paths which are much more elliptical. For instance Finlay's Comet, which has a period of seven years, has its perihelion at a distance only slightly greater than ours, but at aphelion it moves out well beyond mighty Jupiter. And Halley's Comet, with its period of seventy-six years, recedes far beyond Neptune, outermost of the giant planets.

The next diagram (Fig. 5) shows the orbits of the planets of the Solar System, from the Earth to Neptune. (Venus and Mercury, much closer-in than the Earth, cannot be conveniently shown on this scale, while Pluto is the 'odd man out', and for the moment we are entitled to disregard it.) Also shown are the orbits of Finlay's Comet, Halley's Comet, and part of the orbit of Kohoutek's Comet of 1973. Both Finlay's Comet and Halley's Comet are periodical, so that we always know when and where to expect them, even though we can see them only when they are moving in the inner part of the Solar System (when further out, they are too faint to be seen). But what about Kohoutek?

Here we have a path of entirely different type. The eccentricity is very great, and so at aphelion the comet will recede to a tremendous distance. Moreover, while in the far part of its orbit it moves very slowly, and the result is that it will not come back to perihelion for over 70,000 years. To all intents and purposes, it is a non-periodical comet. We could not

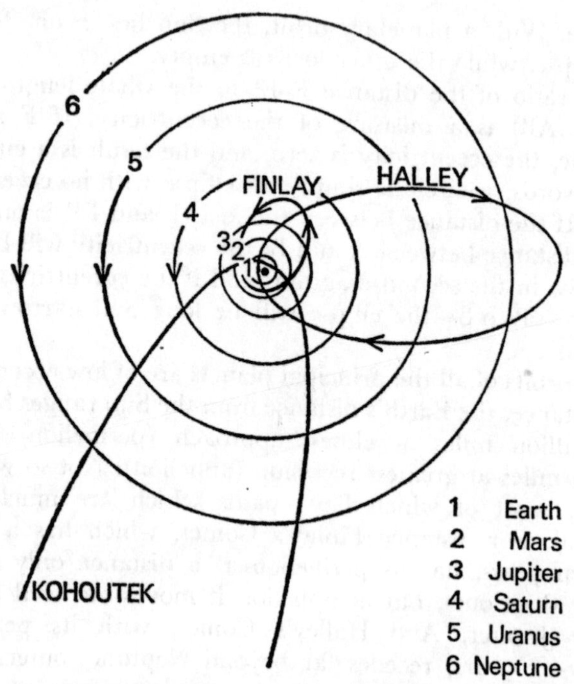


Fig. 5. Orbits of the planets (to scale) and the orbits of Finlay's Comet (short period), Halley's Comet (longer period), and Kohoutek's Comet (non-periodic)

predict it, and this also applies to all other bright naked-eye comets, with the single exception of Halley's. Our more brilliant visitors arrive unannounced, and are apt to take us by surprise.

In an extreme case, a comet may move in an open curve—a path of the kind known as a parabola (Fig. 6). Obviously, a comet moving in a parabolic path will never come back to the Sun at all, and will continue moving outward indefinitely. There are also curves which are still more open, and are known as hyperbolæ. I do not propose to discuss these various curves in any detail, but I must say a little about hyperbolæ, because the whole matter is vital in any discussion of the origin of comets.

If a comet moves in a hyperbolic orbit, it must either have been strongly affected by the gravitational pull of some planet (usually Jupiter), or else approached the Solar System with a

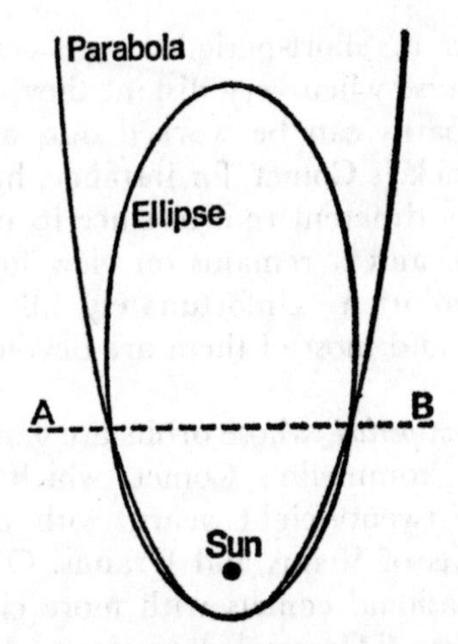


Fig. 6. An ellipse and a parabola. The ellipse is a closed curve, while the parabola is open. When only a portion of the orbit of a comet can be observed (below the line AB in the diagram) it is very difficult to tell whether the comet is moving in a very eccentric ellipse or in a parabola

velocity of its own: that is to say, come from interstellar space. Few comets have hyperbolic orbits, and it is now almost unanimously believed that the former explanation is the correct one. Comets are genuine members of the Solar System. If the orbit becomes 'open', the cause is to be found in what are called planetary perturbations.

Remember, a comet is a very flimsy thing, and it is at the mercy of the gravitational pulls of the planets. If a comet is unwise enough to approach a planet too closely, its orbit will be violently twisted. This has been found to happen on quite a number of occasions. Jupiter, much the most massive of the planets, is particularly powerful as a 'comet disturber', and in an extreme case astronomers believe that a comet may be thrown out of the Solar System altogether.

To sum up, then, comets may be divided into several classes, depending on their orbits:

*1. Short-period comets.* These have periods of between 3·3 years (Encke's Comet) and around twenty years. Many of them have their aphelia (greatest distances from the Sun) at about the mean distance of Jupiter, which again demonstrates how influential Jupiter is. Short-period comets cannot be followed all the time, because when very distant they are too dim to be seen, but their paths can be worked out, and we can keep track of them. Encke's Comet, for instance, has now been seen at more than fifty different returns since its original discovery in the year 1786, and it remains on view for several months during each revolution. Unfortunately all the short-period comets are faint, and most of them are devoid of tails.

2. *Comets of medium period*, whose orbits are generally larger and more eccentric. Crommelin's Comet, which is fairly typical, has a period of twenty-eight years, with a distance range between the orbits of Venus and Uranus. Of rather different type are the occasional comets with more circular paths; for instance, the orbit of Comet Schwassmann-Wachmann 1 lies wholly between those of Jupiter and Saturn, so that it is never out of view on the score of sheer distance.

3. *Long-period comets*, with periods of from about sixty years to as much as 164 years. Comet Grigg-Mellish, which has the longest period of any comet to have been seen at more than one return, can come within the orbit of the Earth, but at aphelion it recedes to almost twice the distance of Neptune. Not many long-period comets have been seen at more than one return, apart of course from Halley's.

4. *Non-periodical comets*. This class includes all the really brilliant comets as well as many fainter ones. Strictly speaking, the term is misleading, because most of the so-called non-periodical comets move in elliptical paths; but because they come to perihelion only once in many centuries (or, in some cases, once in many millions of years) they cannot be predicted. Obviously it is difficult to tell whether a comet of this kind is moving in an eccentric ellipse or in an open curve (a parabola or hyperbola), and calculations of periods are bound to be very arbitrary. Kohoutek's Comet of 1973 is estimated to have a period of about 75,000 years. That of Delavan's Comet of 1914 was calculated to be 24 million years, but all we can really say is that we will not see it again for an extremely long time, if at all!

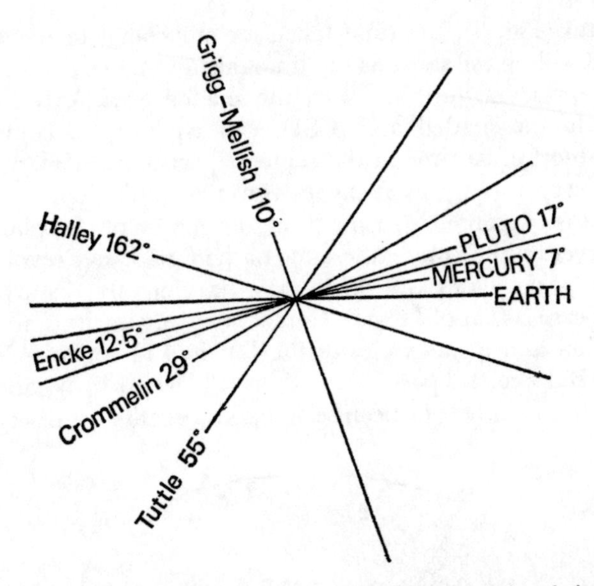


Fig. 7. Inclinations of the orbits of some comets and planets, in relation to the plane of the Earth's orbit. The inclinations are less than 4 degrees for all the planets except for Mercury and Pluto. Comets may have orbits of very high inclination. Of those shown in the diagram, Halley and Grigg-Mellish move in a retrograde direction

Clearly, then, most cometary orbits are very different from those of the planets insofar as eccentricity is concerned. Moreover, the planets have paths which lie in much the same plane. Only Pluto has an inclination of more than 8 degrees with respect to the orbit of the Earth, so that if we draw a plan of the Solar System on a flat piece of paper we are not very far wrong. Comets do not behave in the same manner, and the inclinations may be of any value – 29 degrees for Crommelin's Comet, 12½ degrees for Encke's, and so on, as shown in Figure 7. There are also various comets which travel in a wrong-way or retrograde direction as compared with the planets. Halley's Comet belongs to the retrograde class.

Having discussed Kepler's First Law—a planet (or a comet) moves in an elliptical orbit, with the Sun in one of the foci—let us turn to the Second. In plain language, this Law states that a body moving round the Sun will travel quickest when it is closest-in, slowest when it is at its most remote. In the diagram (Fig. 8), the time taken for our comet to move from A to B will be the same as the time taken to move from C to D. If S represents the Sun, then the shaded area ASB must be equal to the shaded area CSD. Consequently, a comet will spend most of its time in the remoter part of its orbit, and will move very rapidly as it passes through perihelion.

Because a comet is so much at the mercy of the planets, it will never follow exactly the same path in successive revolutions, and in some cases the orbit may be violently disturbed. A classic case is that of Lexell's Comet of 1770, which came within a million and a half miles of the Earth, and was visible with the naked eye. Its period was then calculated to be about $5\frac{1}{2}$ years, but it has never been seen again, because it subsequently

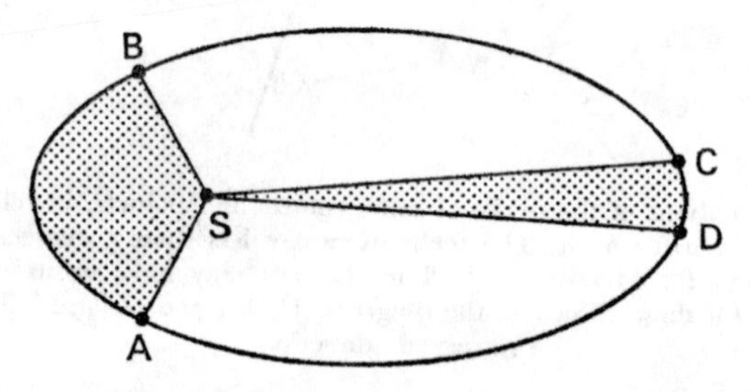


Fig. 8. Kepler's second law. The object moves most quickly when it is nearest to the sun, so it moves from A to B in the same time that it takes to move from C to D. The area ASB is equal to the area CSD

made a close approach to Jupiter—and in its new orbit, it stays so far away from the Earth that nobody has ever been able to track it down. Another naked-eye comet, discovered by the Italian astronomer Francesco di Vico in 1864, has suffered several series of perturbations by Jupiter, and at present it does not come close enough to us to be visible without a telescope. Between 1894 and 1965 it 'went missing', and its recovery in the latter year was very much of a mathematical triumph.

Astronomers often speak of Jupiter's 'comet family', since so many short-period comets have their aphelia at approximately the distance of Jupiter from the Sun (rather less than 500 million miles on average). Originally it was thought that comets came from interstellar space and were literally captured by Jupiter, so that they were forced to remain members of the inner Solar System. This intriguing idea has long since been consigned to the scientific scrap-heap, and it is not certain whether even Jupiter is capable of expelling a comet permanently from the Sun's family, but the influence of the Giant Planet is obvious enough. Alleged comet families associated with the other giants (Saturn, Uranus and Neptune) are nowadays regarded as highly dubious.

To calculate a comet's orbit is no easy task, though it is true that nowadays a great deal of really useful work in this field is carried out by talented amateur mathematicians. The first predicted return was that of Halley's Comet, in 1758–59. Previously, all comets had been regarded as cosmic strays, and even Kepler had believed them to travel in straight lines through space rather than obey the Laws which he had drawn up for the planets.

Finally, there is the question of comet nomenclature. In most cases a comet is named in honour of its discoverer; thus D'Arrest's periodical comet was first seen (in 1851) by the German astronomer Heinrich D'Arrest, and the notoriously disappointing Kohoutek's Comet of 1973 was detected by Dr. Lubos Kohoutek at the Hamburg Observatory. If a comet-hunter has more than one discovery of a periodical comet to his credit, a number is added after his name to show the sequence; thus Ernst Wilhelm Tempel found two, still known as Tempel 1 and Tempel 2. Sometimes the name commemorates the mathematician who first computed the orbit; such are the comets of Encke, Lexell, Crommelin and—of course—Halley. Where two or more astronomers are independent discoverers, either at the same apparition or at different ones, their names may be bracketed; this explains Schwassmann-Wachmann, Grigg-Mellish, Giacobini-Zinner, Arend-Roland, Tuttle-Giacobini-Kresák and others. When a comet is first seen, it is given a letter; thus the first comet to be discovered in 1977 became 1977a, the second 1977b and so on. A more permanent designation follows the date of perihelion, so that the first comet to reach perihelion in 1977 became 1977 I and the second 1977 II. It is seldom that a year passes by without several new comet discoveries, quite apart from the returns of the periodical comets which we can predict with pleasing confidence.

*Chapter Four*

# THE COMET-HUNTERS

MODERN ASTRONOMY HAS BECOME highly specialized. This is true not only in the professional field, but also among the amateurs. There are some enthusiastic observers who concentrate upon studying the Sun; others (such as myself) are interested mainly in the Moon and planets, while still others are concerned with variable stars. Comet-hunters, generally speaking, are a race apart. They do not need powerful telescopes; what they must have is a wide field of view and only moderate magnification. Powerful, well-mounted binoculars are ideal, always provided that the observer has a really encyclopædic knowledge of the sky.

Of course, the periodical comets—usually very faint—are almost always recovered by professional workers, and there are some specialists, such as Miss Elizabeth Roemer in the United States, who are pre-eminent in this research. But the non-periodical comets can appear in any part of the sky at any time, and few professionals feel inclined to spend hour after hour scanning the heavens in the hope of picking up a new visitor. This is why the amateur's work is so valuable. Now and then there is what may be called an accidental discovery. A friend of mine once discovered a comet when he was testing a boy's home-made spectacle-lens telescope; he pointed the telescope out of the bedroom window, focused it, and—lo and behold!—there was an unknown comet. But in the ordinary way, comet discoveries are made as a result of protracted and painstaking searches.

Looking back over astronomical history, there are some great names in the story of comets; and none is more honoured than that of Charles Messier, even though he is now best remembered for something quite different. He was born in Lorraine in 1730, and spent most of his active life in Paris. His main interest was in comet-hunting, and altogether he discovered more than a dozen, but for some time he found that he was being plagued by objects which did not interest him in the least. These were the star-clusters and nebulæ.

A comet, as we have noted, often appears in the guise of a faint luminous patch against the background of stars. The same is true of star-clusters, which are exactly what their name implies—groups of associated stars. Some clusters are easy naked-eye objects, and there can be few amateurs who do not know the Pleiades or Seven Sisters in the constellation of Taurus. Other naked-eye clusters are the Hyades, also in Taurus, round the bright orange-red star Aldebaran; Præsepe, the 'Beehive' in Cancer; and, in the southern hemisphere, the glorious 'Jewel Box', Kappa Crucis in the Southern Cross. Individual stars are very obvious in groups such as these. Yet with fainter, telescopic clusters the separate stars are much less easy to make out, and it is only too easy to confuse a cluster with a comet.

The same applies to nebulæ and galaxies. A nebula is a mass of dust and gas, shining because of stars in or close by it; a galaxy is a separate star-system, perhaps many millions of light-years from us, and well outside the Milky Way galaxy in which we live.

As he searched the sky, Charles Messier continually picked up clusters and nebulæ. Eventually he lost patience with them. To check each misty object wasted an enormous amount of time, and he badly needed a catalogue to which he could refer. Since no such catalogue existed, Messier decided to compile one; and he did. In 1781 he published a list of over a hundred clusters, nebulæ and galaxies, giving a number to each one; thus the celebrated Crab Nebula in Taurus became Messier 1 (M.1), the Orion Nebula M.42, the Andromeda Galaxy M.31, the Pleiades M.45 and so on.

Certainly this made his searches much easier, but the sequel was ironical. Though Messier discovered many comets, it so happened that none of them proved to be particularly bright or particularly interesting, so that they are remembered today only by a relative handful of enthusiasts. On the other hand, his catalogue of clusters and nebulæ became a classic, and the M numbers are still used by astronomers all over the world.

Messier continued his comet-hunting until late in his life (he died in 1817). He had, of course, equally enthusiastic contemporaries, such as Pierre Méchain, also of the Paris Observatory, who discovered eight comets between 1781 and 1799. There was also Caroline Herschel, the first and perhaps the most celebrated of all women astronomers. She was the sister of William Herschel, discoverer of the planet Uranus and one of the greatest observers of all time. Herschel undertook systematic 'reviews of the heavens' with telescopes he had made himself, and Caroline was his faithful assistant; for night after night she would stay with him, checking and recording as he carried out his work. Not content with this, she conducted comet-hunts on her own account, and managed to make six independent discoveries.

Another indefatigable comet-hunter, Jean Louis Pons, had a most unusual career. In 1789, at the age of twenty-eight, he was appointed to a post at the Marseilles Observatory—as doorkeeper and general handyman! Successive directors took an interest in him, and Pons began to hunt for new comets. Altogether he discovered thirty-seven, which was an amazing feat by any standards. Needless to say, he did not remain a doorkeeper; he rose to the rank of Director, first at the Marlia Observatory at Lucca and then at the Observatory of Florence.

Much later there came three American comet-hunters: W. L. Brooks, Lewis Swift, and above all Edward Emerson Barnard, who was renowned for his keen eyesight.* When in his early twenties, in 1881, Barnard discovered a comet which subsequently became brilliant, and he continued to search during most of the rest of his long career, though his most valuable work was in the field of stellar astronomy. Today we have men of the calibre of Leslie Peltier in America, Charles Bradfield in Australia, George Alcock in England and Jack Bennett in South Africa, to say nothing of the enthusiastic team of comet-searchers in Japan. Alcock, a schoolmaster in Peterborough, now has four comet discoveries to his credit, as well as four novæ or new stars. (Actually, the name 'nova' is misleading; the object is not a new star at all. What happens is that a

* It seems quite definite that when using one of the largest refractors in the world, Barnard actually saw craters on Mars—thereby anticipating the Mariner and Viking probes by over seventy years. He did not publish his observations of the Martian craters, because he was convinced that nobody would believe him. I have told the story—so far as anybody knows it—in my book *Guide to Mars* (Lutterworth Press, 1977).

formerly faint star flares up to many times its normal brilliancy before fading back to obscurity. Alcock's first nova discovery was made in 1967, when he detected a particularly interesting star in the constellation of the Dolphin; it is now known as HR Delphini. It remained visible with the naked eye for almost a year, and even today, in 1977, it remains within the reach of comparatively small telescopes.)

Alcock does not have a telescope. Instead, he uses a pair of powerful, firmly-mounted binoculars. He knows the sky rather better than the back of his hand, so that he can detect any new arrival at a glance. At one period during his hunting he found two new comets within a week!

In the southern hemisphere there is Jack Bennett, who lives just outside Pretoria, and observes with a portable, wide-field telescope which was originally designed for following artificial satellites. So far he has found two comets, plus several others which had been detected slightly earlier without his knowledge, and several more which were not confirmed elsewhere, so that they go down in the records as 'comets which got away'. The first Bennett's Comet was that of 1970, which was one of the brightest of modern times. The second was much fainter, and failed to come up to expectations—which was hardly the fault of the discoverer. Bennett is also the living astronomer who has made a visual discovery of a supernova, which is a Titanic stellar outburst in a far-away galaxy. At the time Bennett was not looking for supernovæ; he was on the track of new comets, but as soon as the galaxy came into the field of his telescope he realized that, in his own words, 'there was something wrong with it'.

How does one set about searching for comets? Haphazard sweeping is of no use at all; the chances against success are enormous. What has to be done is to select an area of the sky, and use optical equipment to scan to and fro, noting the familiar stars and checking to see whether anything unusual has made its appearance. I have said that one's knowledge of the sky needs to be encyclopædic, and this is certainly true. Alcock, for example, spent years in memorizing the positions and characteristics of some 30,000 stars, so that it is not surprising that few newcomers will escape him. This is something that few people master really adequately. I have carried out a good deal of lunar and planetary work, but I am by no means competent to search for comets, as I know very well.

Even when a faint comet is found, it is not always easy to identify, and I can cite a personal case here. Some years ago I was at the Observatory of Armagh, in Northern Ireland, which is equipped with a 10-inch refractor (that is to say, a telescope with its object-glass ten inches in diameter). We had a report that a faint comet had been found, and we were officially asked to confirm it, so I went out to check. I turned the telescope to the indicated position, using the widest field that I could manage, and found—nothing. Evidently the position we had been given was wrong, but presumably it was not likely to be far in error, so I began searching.

Again the result was negative, and I was beginning to lose all confidence when I suddenly came upon a very dim patch of light. It could have been a cluster of a nebula, but unlike Messier, so long before, I had adequate catalogues, and a quick search failed to reveal any nebulous object in the right position. Presumably, then, the faint glow must be the comet, but the only way to find out was to watch it until I could detect some movement. After half an hour I was satisfied. The patch had shifted, slowly but perceptibly, against the background stars, and I felt justified in sending off a telegram of confirmation. Undoubtedly it was the comet, but I am the first to admit that I would never have noticed it had I not been given all the relevant information.

Today, it seldom happens that a comet escapes detection before it becomes bright. Most of the major comets are found when they are still a long way from the Sun and the Earth, though there are some which are badly placed and so pass through perihelion without being noticed. Such was Van den Bergh's Comet of 1974, which had an exceptionally great perihelion distance—about 560,000,000 miles, so that it never came within the orbit of Jupiter, and although it was large by cometary standards it never became bright. The Canadian astronomer Sidney van den Bergh discovered it, and when the orbit was worked out it was found that perihelion had been passed some time earlier—on 14 August 1974. This 'distance record' has since been broken by Schuster's Comet of 1976, discovered on 25 February of that year by Hans

Schuster, using a 39-inch Schmidt telescope at the European Southern Observatory in Chili. The perihelion distance of Schuster's Comet was 639,000,000 miles, not far from midway between the orbits of Jupiter and Saturn. I suspect that this record is likely to stand for a long time.

The most distant comet ever observed was followed out to a distance of over 1,000,000,000 miles, so that it was then between the orbits of Saturn and Uranus. This was Stearns' Comet of 1927, which was under observation for over four years; had it had a more normal orbit it would have become brilliant, but its minimum distance from the Sun was well over 300,000,000 miles.

Once a comet has been found, the essential needs are to measure its position with respect to the background stars, and to find out how it is moving. Only when it has been under observation for some time can mathematicians work out a reliable orbit. Here, too, amateurs can play a useful rôle by taking photographs of the comet in its star-fields. Neither must we forget the pure theorists, who examine the observations and work on the calculations. Modern computing machines make the task much easier, but the skill needed is as vital as it has ever been.

I do not propose to say much more about comet-hunting here, simply because it is something which I have never tried myself—except on two occasions, each time with predictably negative results. Occasionally the Moon passes directly in front of the Sun, blotting out the solar disk for a few minutes and producing a total eclipse. The sight is magnificent, because while the Moon is acting as a screen one can see the Sun's atmosphere—the red chromosphere, which is a layer of gas above the brilliant surface; the so-called prominences, once (misleadingly) called Red Flames, also made up of glowing gas; and the superb pearly corona, the Sun's outer atmosphere, which stretches across the sky. As soon as the last segment of the bright dusk of the Sun is hidden, all Nature seems to come to a stop. The sky darkens, and the stars and planets shine out.

Obviously, this is the only time when we can see a dark sky with the Sun above the horizon. In 1882, when a total eclipse was visible from Egypt, photographs showed not only the corona, but also a bright comet quite close to the Sun. It had never been seen before, and it was never seen again, so that this is our only record of it. It is commonly known as Tewfik's Comet, in honour of the then ruler of Egypt.

In 1882 photography was in its infancy, astronomically speaking. Should a brilliant comet be seen during a modern eclipse, it would be widely recorded. However, there is always the chance that a fainter one would escape notice, because almost everybody would be concentrating upon the Sun itself rather than on surrounding areas of the sky. During two of the four total eclipses that I have seen, I have made elaborate plans for taking sky photographs in the faint hope of catching an unknown comet. My first attempt, in Siberia in 1968, was doomed to failure from the outset; totality lasted for only thirty-seven seconds, and the sky never became really dark. I hoped for better luck during the eclipse of 30 June 1973, which I observed from a ship, the MS *Monte Umbe*, twenty-four miles off the coast of Mauritania, in Africa. Totality extended over more than six minutes, which was practically a record; but unfortunately there was a certain amount of high-altitude haze, and I could see no stars at all, though the planets Venus and Saturn shone out. Under such conditions it was clearly pointless to search for faint comets, and I contented myself with carrying out a commentary for B.B.C. television and taking pictures of the corona. At the next eclipse I am fortunate enough to see, I will try again.

Meanwhile, the band of enthusiastic comet-hunters will continue nightly searches. Generally the results will be negative; but if the observer persists for long enough—well, sooner or later he has at least a chance of discovering a comet. And if so, then the sight of the tiny, pale blur of light against the backcloth of remote stars will be full reward for the hundreds of hours of patient work.

*Chapter Five*

# ENCKE'S COMET—AND OTHERS

QUITE APART FROM NEW comet discoveries, several old friends return to perihelion each year. Few periodical comets can be followed throughout their orbits. The most famous example is Comet Schwassmann-Wachmann 1, which keeps strictly to the zone between Jupiter and Saturn; another is Gunn's Comet, first seen in 1969 and whose distance from the Sun ranges from about 230,000,000 miles to 330,000,000 miles, putting it in the asteroid zone; and the more recently discovered Smirnova-Chernykh, which also moves in the asteroid zone. Most comets, however, remain visible for only limited periods, while moving in the inner part of the Solar System. When the comet recedes again, we lose it until it swings back into our range.

Over forty short-period comets have now been observed at more than one return, so that their orbits are well known. Some of them are familiar indeed; Encke's Comet, first seen in 1786 by Méchain, made its fifty-first appearance in 1976. To describe all the short-period comets would be distinctly tedious, so I propose to confine my account of them to comets which have been under observation at some time or other since 1973. I begin with 1973 because it so happened that during that year several comets of special interest were on view.

The names given to the comets are self-explanatory, and almost always follow that of the discoverer or discoverers. (There are a few exceptions, Encke's being one.) The date of perihelion can be calculated accurately, and so can the period, though remember that because a comet is so easily pulled out of its path by the gravitational actions of the planets, it never follows exactly the same orbit twice. The column headed 'magnitude' does, however, require some explanation.

With the stars, and with the planets, magnitude is a measure of apparent brightness. The scale works in the manner of a golf handicap, with the more brilliant performers having the lower values. Stars of magnitude 6 are just visible with the naked eye on a clear night; those of 5 are brighter, 4 brighter still, and so on. Really conspicuous stars, such as Altair in the Eagle, are of the first magnitude, and we can even have zero or negative values. Sirius, the most brilliant star, is of magnitude −1·4. Of the planets, Venus can exceed −4; on the same scale the full moon is about −12, and the Sun is −26. Toward the faint end of the scale, binoculars can show stars down to about magnitude +9; the telescope in my own observatory will carry me down to +15, and the world's greatest telescopes can photograph stars as dim as magnitude +26.

The scale is satisfactory for stars, which are virtually point sources, and for planets, which show very small disks. When we consider the magnitude of a comet, we are bound to be less accurate, because a comet appears as a blur rather than a sharp point. The values given in the following tables refer to the nucleus only, so that the comets may be rather more conspicuous than their magnitudes indicate—though not many of them come within the range of the average amateur-owned telescope.

Here, below, is the list of short-period comets on view during 1973. Three had already been recovered by the start of the year; the rest followed in due course. I have omitted Comet Smirnova-Chernykh, which, although one of those comets which can be followed throughout its orbit, was not discovered until 1975.

| *Name* | *Date of perihelion* | *Period years* | *Max. mag., 1973* |
|---|---|---|---|
| Giacobini-Zinner | 1972 Aug. 4 | 6·52 | 17 |
| Swift-Gehrels | 1972 Aug. 31 | 9·23 | 15 |
| Tempel 2 | 1972 Nov. 15 | 5·26 | 18 |
| Kwerns-Kwee | 1972 Nov. 29 | 9·01 | 16 |
| Gehrels 1 | 1973 Jan. 25 | 14·54 | 19 |
| Reinmuth 1 | 1973 Mar. 21 | 7·63 | 17 |
| Clark | 1973 May 25 | 5·50 | 13 |
| Tuttle-Giacobini-Kresák | 1973 May 29 | 5·56 | 14 |
| Wild | 1973 July 2 | 13·29 | 19 |
| Gehrels 2 | 1973 Dec. 1 | 7·43 | 15 |
| Brooks 2 | 1974 Jan. 4 | 6·88 | 17 |
| Encke | 1974 Apr. 28 | 3·30 | 18 |

| | | | |
|---|---|---|---|
| Reinmuth 2 | 1974 May 7 | 6·73 | 19 |
| Borrelly | 1974 May 12 | 6·77 | 18 |
| Schwassmann-Wachmann 2 | 1974 Sept. 12 | 6·51 | 18 |
| Schwassmann-Wachmann 1 | — | 16·3 | 18 |
| Gunn | — | 6·8 | 20 |

Because Encke's Comet has the longest history, and because it is the brightest of the objects in our list (during 1974 it became a relatively easy object), it is worthy to be discussed first. So far as we are concerned, its story began on 17 January 1786, when Pierre Méchain, Messier's contemporary and countryman, discovered a telescopic comet in the constellation of Aquarius, the Water-bearer. Méchain immediately sent the news on to Messier, who confirmed the discovery on the first clear night after he received the message. The nucleus was fairly bright, but there was no tail. Only a few observations of the comet were made, so that no reliable orbit could be worked out.

On 7 November 1795, Caroline Herschel, presumably taking time off from acting as her brother William's assistant, discovered a comet which was circular in shape, and was not well defined. Other observers also saw it, and were puzzled by the way in which it was moving. Remember, at this time there was only one comet—Halley's—known to travel round the Sun in an ellipse.

Next, on 19 October 1805, the French astronomer Thulis, at Marseilles, found a comet which was just visible with the naked eye. By early November it had developed a short tail, and Johann Encke, Director of the Berlin Observatory, tried to work out an orbit for it. He suggested that it might have a period of about twelve years, but he was the first to admit that his estimate was very uncertain.

The next chapter in the story began on 26 November 1818, when the indefatigable Pons discovered a small, inconspicuous comet without a tail. It remained on view for almost two months, and many careful measurements were made as it tracked slowly against the starry background. Encke set to work—and found that no open-curve (parabolic) orbit would fit the observations. Using a new mathematical method, he calculated that the period could be no more than $3\frac{1}{2}$ years.

He then consulted the comet catalogues, and found that the orbit was strikingly similar to those of Méchain's Comet of 1786, Caroline Herschel's of 1795, and Thulis' of 1805. The identifications could not be regarded as certain, because the earlier comets had been less thoroughly observed (particularly in the case of Méchain's), but Encke was confident enough to predict that there would be another return to perihelion on 24 May 1822.

Astronomers watched. On 2 June 1822 an astronomer named Rumker, in Australia, found the comet just where Encke had said it would be, and calculations showed that it really had passed perihelion toward the end of May. Encke at once worked out the time for the next return: September 1825. Again the comet appeared on schedule. Since then it has been

Fig. 9. Encke's Comet 30 November 1828.

Drawing by W. Struve

missed at only one return: that of 1944, when it was badly placed and when many astronomers had their minds on other things. It is surely fitting to call it Encke's Comet.

When Thulis saw it, in 1805, it was visible with the naked eye. At the 1828 return it reached the fifth magnitude (Fig. 9), and when it appeared in 1871 it developed an interesting, fan-like tail. Nowadays it is seldom so prominent as this, and I have never personally seen it with the naked eye; neither have I seen a definite tail, though I have looked at it quite often since the 1930s. This may not be significant, since I would not for one moment claim to be a skilled comet observer, but there are other suggestions that it is not so bright as it used to be. If so, why not?

Each time a comet comes back to the region of the Sun, it

loses a little of its substance, because the ices in it tend to evaporate. It is estimated that Encke's Comet, which has the shortest period of all and which passes through perihelion thirty times in every century, loses 1/200 of its mass at each return. This wastage is quite appreciable, particularly for a flimsy object such as a comet, and it may well be that our old friend will have a limited lifetime. If it can last for another two hundred returns, it will have been identifiable as a comet for 660 years or thereabouts. We have known it now for almost two centuries, and we may be watching its gradual demise. There have even been suggestions that it will fade away at some time between 1990 and 2000. Personally, I hope that this is not so; the Solar System would seem incomplete without it—and yet it cannot persist for very long on the cosmical time-scale. In the foreseeable future it must die, as a few former comets, such as Biela's, have already been seen to die.

Yet the evidence of fading has been challenged. As will be seen from the 1974 table on p. 53, the maximum magnitude in this year was 4·4; but this is somewhat misleading, and is unconfirmed because when the comet was at its brightest it was hopelessly near the Sun in the sky. All we can really say is that there is a distinct chance that the comet is declining in brilliancy over the years.

Another interesting fact is that the period of Encke's Comet has shortened appreciably. The distance from the Sun ranges between 31½ million miles and 381 million miles, so that it travels from near the orbit of Mercury right out into the asteroid belt (Fig. 10). (It can sometimes approach Mercury within four million miles.) The period shortened by more than a day between 1822 and 1858, and Encke himself was forced to the conclusion that when near the Sun the comet was having to 'push through' some resisting medium (that is to say, thin gas), so that it was braked. In fact, this is not the right answer; the process of evaporation is responsible. The effect is particularly noticeable with Encke's Comet because it comes back to the Sun so often.

Whether it will continue to hold the 'short-period record' remains to be seen. Comet Wilson-Harrington, discovered in 1949, was calculated to have a period of only 2·3 years, with

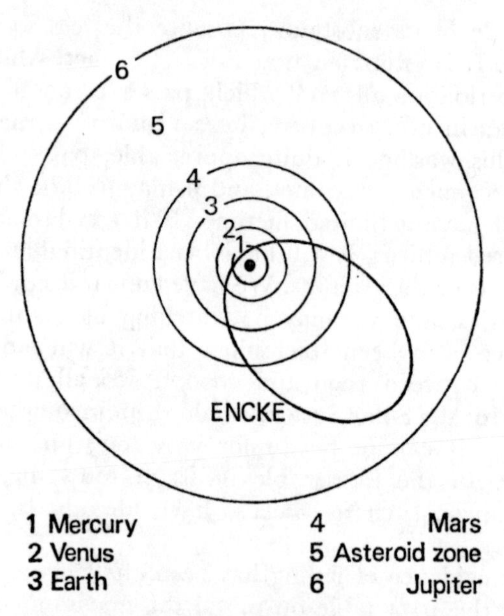


Fig. 10. The orbit of Encke's Comet. At perihelion it is closer to the sun than is Mercury; at aphelion it penetrates the asteroid belt

an orbit taking it from 95,600,000 miles out to 230,000,000 miles from the Sun; but it has never been seen again, and whether it will be recovered in the future is doubtful. At present, then, Encke's Comet stands alone. Its period is more than a year less than that of its nearest rival, Grigg-Skjellerup (5·1 years).

I have spent some time in discussing Encke's Comet because it is of special interest and importance. The other comets in our 1973 list are always fainter than Encke's at its best, but they too have their points of interest, and a few comments about some of them may be worth making here, dim and elusive though the comets themselves undoubtedly are.

Tempel 2 was discovered by Ernst Wilhelm Liebrecht Tempel, a German astronomer living in Italy, as long ago as 1873. (It was his second periodical comet; he had found his first six years earlier.) Its return in November 1972 was the fourteenth to be observed, but it was badly placed, and excessively faint, during 1973.

I. Kohoutek's Comet from Lowell Observatory, 5 and 27 November and 7 December 1973.

II. Kohoutek's Comet from Royal Greenwich Observatory, November 1973.

III. (*top*) Meteor Crater near Winslow, Arizona, 1964. Aerial photography by Patrick Moore.

(*right*) The Leonid Shower, 1833. From an old woodcut.

IV. (*top*) The Andromeda Galaxy (M31), in the centre, looks very much like a comet on this small scale. Photograph by Commander H. R. Hatfield.
(*below*) Comet Kobayashi Berger-Milon, August 1975, centre top. Photograph by D. G. Daniels.

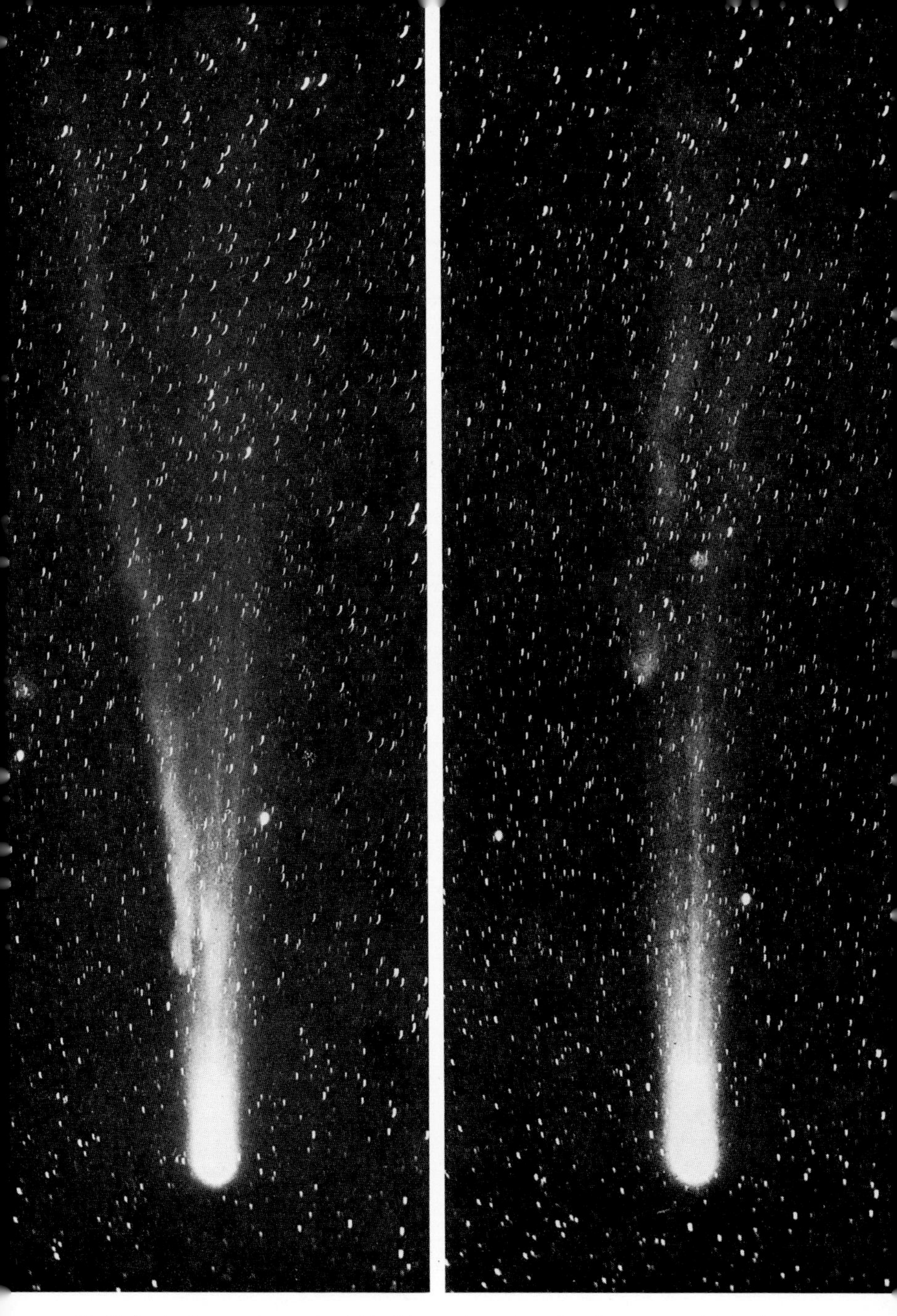

V. Halley's Comet, from the Lick Observatory, 6 and 7 June 1910. Note the changes in the tail between the two photographs.

VI. De Chéseaux's Comet of 1744 as seen from Lausanne, 8 March 1744.

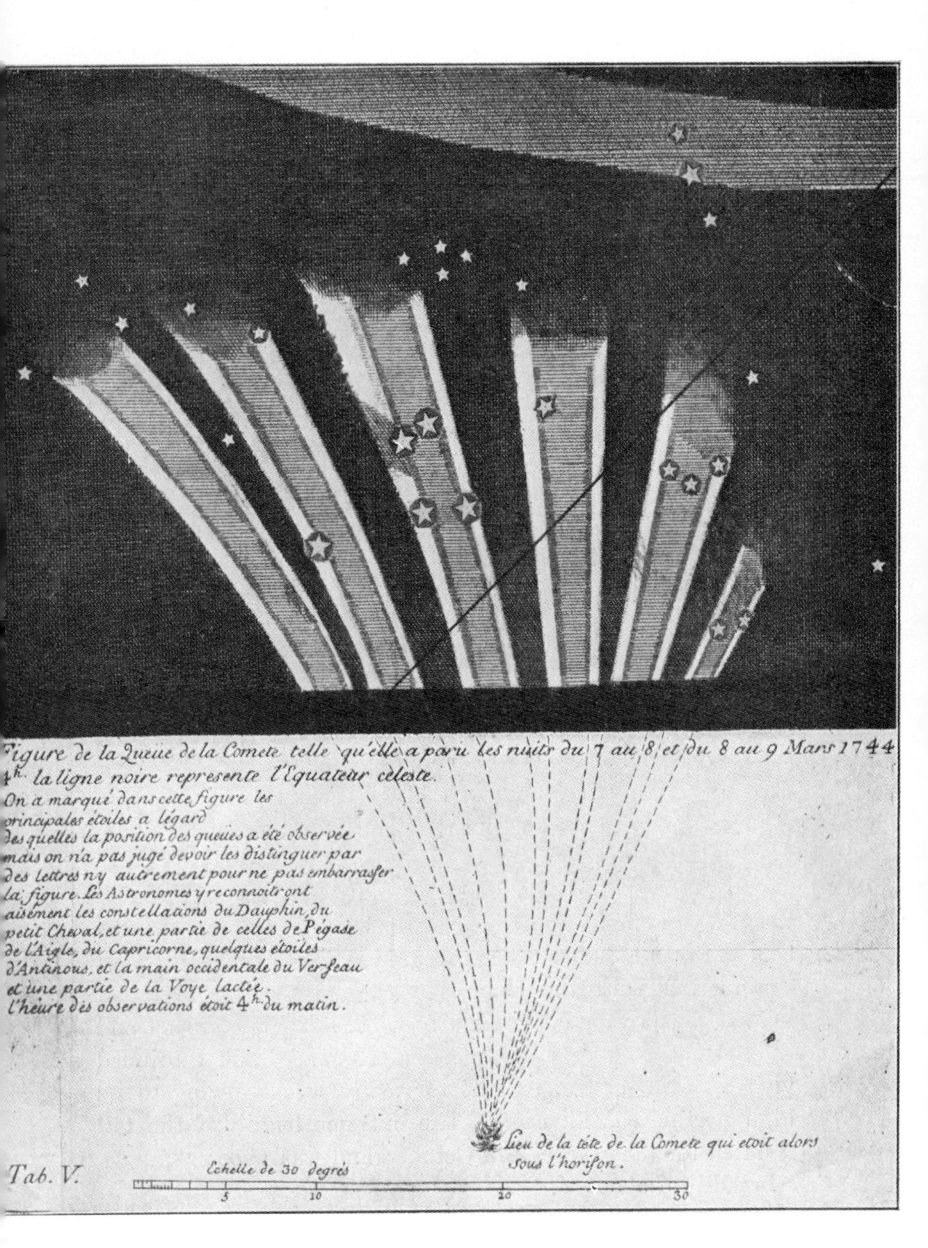


VII. De Chéseaux's drawing of the comet, 7 March 1744.

VIII. (*top*) The Great Comet of 1811. From an old woodcut.
(*below*) Donati's Comet of 1858. From an old woodcut.

IX. A drawing of the Great Comet of 1843 made by an observer at the Cape of Good Hope. According to many observers this was the most brilliant comet of the last century.

X. The Great Comet of 1861 as seen with a 13-inch telescope on July 3. Drawing by W. de la Rue.

XI. The Great Comet of 1882, from the Cape of Good Hope. Photograph by Sir David Gill. This was the first really good photograph of a comet, and so many stars were shown that Gill immediately saw the value of stellar photography.

XII. Comet Brooks 2, photographed from Helwan Observatory with a 30-inch reflector, 25 October 1911.

XIII. (*above*) Comet Arend-Roland, from the Armagh Observatory, 1957. Photograph by E. M. Lindsay.
(*below*) Mrkós' Comet, from Greenwich Observatory, 1957. Photograph by E. A. Whitaker.

XIV. Comet Ikeya-Seki, 1 November 1965. Photograph by A. McClure.

XV. Bennett's Comet of 1970 as seen from England on 4 March.
Photograph by F. J. Acfield.

XVI. (*above*) Biela's Comet, 1846. Drawing by Angelo Secchi.
(*below*) West's Comet, from Bulawayo, Rhodesia, 30 March 1976. Photograph by Jack McBain.

The rather picturesquely-named comet Kwerns-Kwee has a curious history. It was discovered in 1963 during a search for an entirely different periodical comet which had been lost for some time. It was then found that Kwerns-Kwee, the newcomer, had formerly had a period of fifty-one years, and had moved in a path which never brought it anywhere near the Earth, but a close approach to Jupiter in 1961 had altered all this. The orbit had been completely changed, which is why it now becomes bright enough to be detected, though of course it remains very dim. Miss Roemer found it again in July 1971; it was still under observation at the beginning of 1973.

Another 1973 comet which has undergone wild perturbations in its orbit is Brooks 2. It was first seen in July 1889, but its most exciting adventure had occurred three years earlier. What apparently happened was that the comet used to have a period of 29 years, but in March 1886 it had a dramatic encounter with Jupiter, and actually invaded the Giant Planet's system of satellites (Fig. 11). Io, the innermost of the four large Jovian moons, travels round the planet at a distance of only 260,000 miles, and Brooks' Comet passed closer-in than that. In its first orbit it was moving at more than 8 miles per second, but Jupiter 'swung it round' and slowed it down. There were two important results. First, the comet's orbit

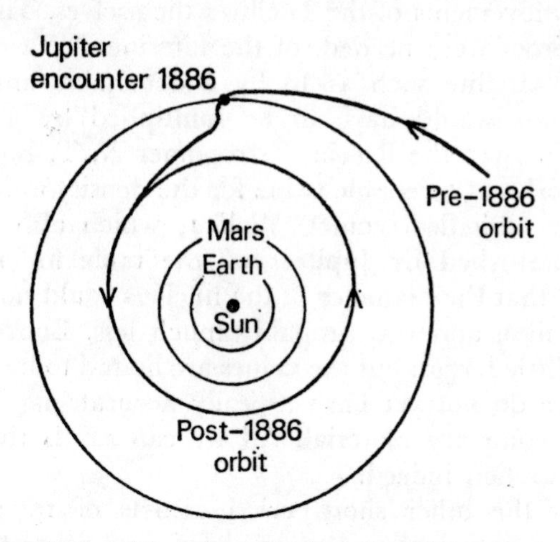


Fig. 11. Changes in the orbit of Comet Brooks 2. In 1886 the comet passed Jupiter and was swung into an orbit of a much shorter period

was changed into a much smaller ellipse with a period of only 6¾ years. Secondly, the comet itself was damaged, because it had been subjected to tremendous gravitational strain. It must also have penetrated the zones of intense radiation surrounding Jupiter, which were studied in detail by the probes Pioneer 10 in 1973 and Pioneer 11 in 1974.

The effects on the comet were still noticeable in 1889, when Edward Emerson Barnard found that it was 'multiple' inasmuch as the main object was attended by four companions —splinters, if you like. Two of the splinters soon vanished; then the third expanded and faded away, and finally the fourth lost its tail and faded out in similar fashion. There was no doubt that the comet had been in grave danger of total disruption, and there was a striking precedent—Biela's Comet, to be described below. Astronomers watched eagerly for the return of Brooks 2 in 1896. It returned on schedule, and when first seen, by Javelle at Nice on 20 June, it was single. Since then it has been seen at every return apart from those of 1918 and 1967, when it was missed because it was so badly placed. It is, of course, very faint—but after its devastating experience near Jupiter, it is lucky to have survived at all!

Note, incidentally, that even though it became involved in the Jovian satellite system, Brooks 2 had no measurable effect upon the movements of the satellites themselves. This is extra proof, if proof were needed, of the flimsiness of a comet. To perturb a satellite such as Io by a detectable amount, the comet's mass would have to be multiplied by a factor of millions. In 1947 the Russian astronomer N. T. Bobrovnikov tried to work out a reliable value for the density and diameter of another periodical comet, Wolf 1, which also had been strongly perturbed by Jupiter (*see* the table for 1975). He calculated that the diameter of the nucleus could not be more than 5½ miles, and was probably much less. Encke's Comet may be a little larger, but the values are bound to be arbitrary, because we do not yet have a really accurate figure for the density of cometary material. All we can say is that comets are very rarefied indeed.

Most of the other short-period comets of 1973 may be passed over with no more than brief comment. Giacobini-Zinner is interesting because meteors spread along its orbit can occasionally produce major showers, as in 1933. Borrelly's Comet was discovered in 1904, and has been seen at every subsequent return apart from those of 1939 and 1946. Wild's Comet, with a rather longer period, had been seen for the first time in 1960, so that it was making its first predicted return. On the other hand, Schwassmann-Wachmann 2 has never failed to come under observation at each perihelion passage since 1929. It also has had a chequered history. Before 1921 its orbit was much more circular than is the case today, and its period was over nine years. As usual, Jupiter was the disturbing influence.

Another comet on view during 1973 was Schwassmann-Wachmann 1, discovered in 1925 by two German astronomers working together at the Bergedorf Observatory. Here we have a really exceptional orbit, more like that of a planet than a comet. There were even some early suggestions that it might be an asteroid, but this is certainly not so. It shows a slight but distinct coma, which no asteroid can do, and it is subject to sudden, unpredictable outbursts which bring it within the range of fairly small telescopes.

In its 1973 orbit, the distance from the Sun ranged between about 510,000,000 miles and about 67,000,000 miles; the values for Jupiter and Saturn are, on average, 483,000,000 and 886,000,000 miles respectively. The inclination of the orbit was 9½ degrees, and the period rather over sixteen years. This remained the case until the middle of 1974. In June of that year there was a close approach to Jupiter, and the orbit of the comet was altered (Fig. 12). By 1979 it will be moving in a near-circle, with a distance from the Sun ranging between 521,000,000 miles and 591,000,000 miles.

Generally the magnitude is about 18, but sometimes a star-like point appears in the nucleus and develops quickly; the brightness increases, and the nucleus expands into a small disk. Slowly the size grows, and the brightness drops once more until the comet returns to its normal state. Obviously, material is being sent outwards, and the velocities involved may reach several miles per second.

These remarkable outbursts were first studied in detail by the East German astronomer Nikolaus Richter several decades ago. He found that in September 1941 the magnitude rose

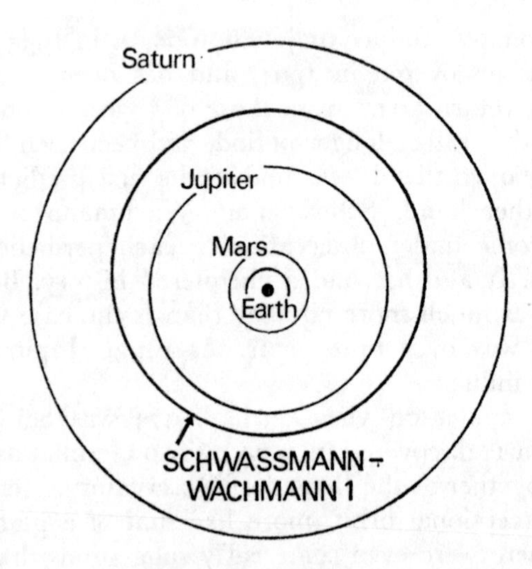


Fig. 12. Orbit of Comet Schwassmann-Wachmann 1. The eccentricity is low, and the comet keeps strictly to the area between the orbits of Jupiter and Saturn, so that it is observable every year

to 11. There was another flare-up in 1946, observed at Yerkes in the United States by the famous comet expert G. van Biesbroeck. The magnitude was normal at the start of the year, but then came the outburst, and by 26 January the magnitude was 9·4, almost within binocular range. This meant that the luminosity had increased some 3,000-fold. Other sudden rises in brightness occurred in 1959, 1961 and 1965. The most recent outburst was that of autumn 1976, when the magnitude rose to well above 12.

It is true that major disturbances have been seen in other comets, but Schwassmann-Wachmann 1 remains in a class of its own, and at the moment we simply do not know exactly what happens. So far as we know, the other comets with orbits of low eccentricity do not act in comparable fashion. Gunn's does not; neither did Oterma's, which used to move in a more or less circular path, though in 1965 it was perturbed by Jupiter and thrown into its modern orbit of greater ellipticity and a period of nineteen years. No outbursts have been seen in Comet Smirnova-Chernykh, though since it was discovered only in 1975 we can never be sure what it will do.

At any rate, Schwassmann-Wachmann 1 can be kept under observation all the time, except when it is too close to the Sun in the sky. Though it is usually much too faint to be detectable with modest equipment, it is useful to keep checks on its position as given in the various annual publications, because an outburst may happen at any time.

Four of the comets in the 1973 list had not been predicted. One was found by Clark in Australia; it proved to have a period of 5½ years, so that it will no doubt be back once more at the end of 1978. The other three were found by the American observer T. Gehrels. Two were unremarkable, but the other turned out to be identical with the long-lost comet originally found by Lewis Swift in 1889. It had made nine returns without being detected, and had been more or less given up. It is now known as Swift-Gehrels.

Let me now give the lists for the period 1974–76. Periodical comets seen for the first time are identified by an asterisk. I have omitted Schwassmann-Wachmann 1.

| *Name* | *Perihelion* | *Period, years* | *Maximum magnitude* |
|---|---|---|---|
| | *1974* | | |
| Brooks 2 | 1974 Jan. 3 | 6·88 | 18 |
| Schwassmann-Wachmann 3 | 1974 Mar. 17 | 5·40 | 18 |
| Du Toit 1 | 1974 Apr. 4 | 14·96 | 17 |
| Encke | 1974 Apr. 29 | 3·30 | 4 |
| Reinmuth 2 | 1974 May 7 | 6·71 | 17 |
| Borrelly | 1974 May 12 | 6·77 | 17 |
| Forbes | 1974 June 2 | 6·40 | 14 |
| Schwassmann-Wachmann 2 | 1974 Sept. 12 | 6·51 | 17 |
| Longmore* | 1974 Nov. 4 | 7·05 | 18 |
| Swift 2 | 1974 Nov. 5 | 7·15 | 18 |
| Wirtanen | 1974 Dec. 20 | 6·67 | 20 |
| Honda-Mrkós-Pajdusáková | 1974 Dec. 28 | 5·28 | 13 |
| Gunn | 1976 Feb. 11 | 6·81 | 16 |
| | *1975* | | |
| Schwassmann-Wachmann 2 | 1974 Sept. 12 | 6·51 | 17 |
| Honda-Mrkós-Pajdusáková | 1974 Dec. 28 | 5·28 | 12 |
| Boethin* | 1975 Jan. 6 | 10·97 | 13 |
| Kohoutek* | 1975 Jan. 18 | 6·20 | 15 |
| West-Kohoutek-Ikemura* | 1975 Feb. 25 | 6·10 | 13 |
| Arend | 1975 May 25 | 7·89 | 19 |

| | *1975* | | |
|---|---|---|---|
| Metcalf | 1975 June 20 | 7·77 | 18 |
| Perrine-Mrkós | 1975 Aug. 2 | 6·78 | 15 |
| Smirnova-Chernykh* | 1975 Aug. 6 | 6·64 | 18 |
| Giacobini | 1975 Nov. 7 | 6·61 | 18 |
| Wolf 1 | 1976 Jan. 25 | 8·42 | 18 |
| Gunn | 1976 Feb. 11 | 6·81 | 16 |
| Harrington-Abell | 1976 Apr. 21 | 7·58 | 19 |
| | *1976* | | |
| Arend | 1975 May 25 | 7·89 | 19 |
| Perrine-Mrkós | 1975 Aug. 2 | 6·78 | 18 |
| Smirnova-Chernykh | 1975 Aug. 6 | 6·64 | 18 |
| Wolf 1 | 1976 Jan. 25 | 8·43 | 19 |
| Gunn | 1976 Feb. 11 | 6·81 | 15 |
| Churyumov-Gerasimenko | 1976 Apr. 12 | 6·60 | 17 |
| Harrington-Abell | 1976 Apr. 21 | 7·58 | 19 |
| Neujmin 2 | 1976 June 18 | 5·40 | 14 |
| D'Arrest | 1976 Aug. 13 | 6·23 | 8 |
| Klemola | 1976 Aug. 20 | 11·0 | 16 |
| Schaumasse | 1976 Sept. 5 | 6·36 | 20 |
| Pone-Winnecke | 1976 Dec. 8 | 6·35 | 20 |
| Johnson | 1977 Jan. 8 | 6·77 | 19 |
| Taylor | 1977 Jan. 23 | 6·95 | 16 |
| Faye | 1977 Feb. 27 | 7·39 | 17 |
| Kopff | 1977 Mar. 8 | 6·43 | 20 |
| Grigg-Skjellerup | 1977 Apr. 11 | 5·10 | 19 |
| Encke | 1977 Aug. 16 | 3·31 | 20 |

The list for 1977 was similar:

| *Name* | *Perihelion* | *Period, years* | *Maximum magnitude* |
|---|---|---|---|
| | *1977* | | |
| Smirnova-Chernykh | 1975 Aug. 6 | 6·64 | 18 |
| Gunn | 1976 Feb. 11 | 6·81 | 18 |
| Churyumov-Gerasimenko | 1976 Apr. 12 | 6·60 | 17 |
| Johnson | 1977 Jan. 8 | 6·77 | 17 |
| Taylor | 1977 Jan. 23 | 6·95 | 16 |
| Faye | 1977 Feb. 27 | 7·39 | 17 |
| Kopff | 1977 Mar. 8 | 6·43 | 17 |
| Gehrels 3* | 1977 Mar. 12 | 8·37 | 17 |
| Grigg-Skjellerup | 1977 Apr. 11 | 5·10 | 13 |
| Encke | 1977 Aug. 16 | 3·31 | 8 |

| | | | |
|---|---|---|---|
| Tempel 1 | 1978 Jan. 11 | 5·50 | 19 |
| Arend-Rigaux | 1978 Feb. 2 | 6·83 | 17 |
| Tempel 2 | 1978 Feb. 20 | 5·27 | 19 |
| Wolf-Harrington | 1978 Mar. 16 | 6·53 | 17 |
| Whipple | 1978 Mar. 27 | 7·44 | 18 |
| Tsuchinshan 1 | 1978 May 10 | 6·60 | 20 |
| Kojima | 1978 May 24 | 7·58 | 18 |
| Ashbrook-Jackson | 1978 Aug. 19 | 7·42 | 19 |
| Comas Solá | 1978 Sept. 24 | 8·93 | 19 |
| Van Biesbroeck | 1978 Dec. 3 | 12·39 | 20 |

Several of these comets had not been seen since the year of their discovery—as long ago as 1906 in the case of Metcalf's Comet, which is believed to have passed near Jupiter in 1911, 1935 and again in 1969, with consequent disturbances in its orbit. Against this there were familiar visitors such as Finlay's Comet, originally seen by the astronomer of that name at the Cape Observatory in 1886, which duly returned in 1974—the thirteenth apparition on record. Faye's Comet, which came back on schedule, was first seen as long ago as 1843.

Yet one thing is obvious: of all the flock, there are very few comets which come within the range of small telescopes. In the lists given here, only the comets of Encke and D'Arrest are not excessively faint. It would be pleasant to have a bright comet every few years, but Nature does not oblige us, and there is only one comet with a period of less than a century which can be relied upon to make a brave showing. This is Halley's Comet, to which we will now turn.

*Chapter Six*

# HALLEY'S COMET

Of all the comets in the sky,
There's none like Comet Halley.
We see it with the naked eye,
And periodically.

I AM NOT SURE who wrote this immortal verse. It was not, I hasten to add, Shakespeare (or Bacon, or Marlowe), and it sounds rather more like McGonagall, but it is the sort of jingle which sticks in one's head. Certainly Halley's Comet is unique; it can become striking, and there is no doubt that it will be found again as it comes in toward its next perihelion in 1986.

On 15 August 1682 a comet was recorded by a German astronomer named Georg Dorffel. It was also seen at the relatively new observatory at Greenwich, and some observations of it were made by the Astronomer Royal, the Rev. John Flamsteed. Another man who looked at it attentively was Edmond Halley,* later to succeed Flamsteed at Greenwich. The comet brightened up steadily, and became really brilliant, with the tail stretching for some distance across the sky. Various drawings of it have survived—one, for instance, by Hevelius of Danzig (the modern Gdańsk), a leading observer of the period. One must admit that the drawings are somewhat peculiar, but they were good enough to show that the comet was very prominent indeed.

At this time, no comet was known to be a member of the Solar System. (Kepler had believed them to travel in straight lines, and not to obey the Laws of Planetary Motion.) Halley was not so sure, and he was of course a close friend of Isaac Newton, who was then working on the theories of gravitation.

* The correct spelling of the name really is Edm*o*nd, and not the often-found Edm*u*nd. My colleague Colin Ronan, the scientific historian, maintains that 'Halley' should rhyme with 'poorly', not 'valley'—which would, alas, ruin the verse with which I introduced this chapter!

Could it be that the 1682 comet was periodical, and had been seen at earlier returns?

Halley was in no hurry. He waited until Newton's new theories had become established, and then he set to work. Using Flamsteed's observations as well as those of other astronomers, he calculated the orbit of the 1682 comet as carefully as he could, and found that the measured positions would fit in with an assumed period of about seventy-five years. He then checked back to see what other comets of similar brilliancy had been seen in the past. Among them were comets which had been under observation for some time in 1607 and in 1531. The intervals between perihelion passages were not exactly equal, but Halley found that between 1607 and 1682 the comet must have passed close enough to Jupiter for its period to be appreciably shortened. Before long he had come to the firm conclusion that these three comets—1531, 1607 and 1682—were one and the same.

His results were communicated to England's senior scientific organization, the Royal Society, in 1706. He wrote in Latin, and the following is a translation of the relevant part of his paper:

> Now many things lead me to believe that the comet of the year 1531, observed by Apian, is the same as that which in the year 1607 was described by Kepler and Longomontanus, and which I myself saw and observed at its return in 1682. All the elements agree, except that there is an inequality in the times of revolution; but this is not so great that it cannot be attributed to physical causes. For example, the motion of Saturn is so disturbed by the other planets, and especially by Jupiter, that its periodic time is uncertain to the extent of several days. How much more liable to such perturbations is a comet which recedes to a distance nearly four times greater than that to Saturn, and a slight increase in whose velocity could change its orbit from an ellipse into a parabola? The identity of these comets is confirmed by the fact that in 1476 a comet was seen, which passed in a retrograde direction between the Earth and the Sun, in nearly the same manner; and although it was not observed astronomically, yet from its period and its path I infer that it was the same comet as that of the years 1531, 1607 and 1682. I may, therefore, with confidence predict its return in the year 1758. If this prediction is fulfilled, there is no reason to doubt that other comets will return.

He added, modestly, that if he were proved right, posterity would not fail to acknowledge that the discovery had first been made by an Englishman.

Halley could not hope to live to see his prediction fulfilled; he died in 1742. Later, some new calculations were made by three French astronomers, Lalande, Clairaut and Madame Lepaute. In November 1758—when Messier, among others, had already begun to search—Clairaut announced that Saturn would delay the comet for 100 days and Jupiter for 518 days, so that the actual date of perihelion would be in the spring of 1759. Yet Halley was vindicated in every respect. On Christmas Night 1758, a German amateur named Johann Palitzsch duly picked up the comet; Messier saw it on 21 January 1759, and it passed through perihelion on 12 March. Extensive observations of it were made by astronomers all over the world.

By common consent Halley's name was attached to the comet, and surely the honour was well deserved. Remember, his investigation was something entirely new, and it showed,

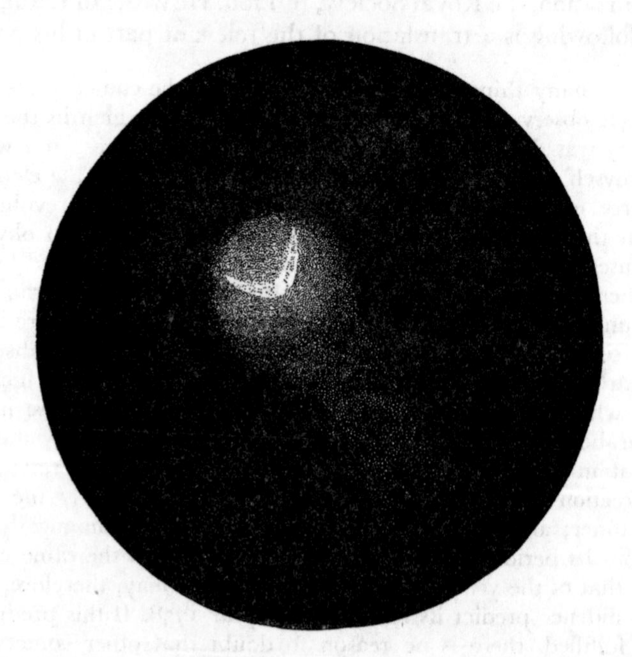

Fig. 13. Halley's Comet, 11 October 1835. Drawing by C. Piazzi Smyth

for the first time, that some comets at least had to be ranked as true members of the Solar System.

The next return was due in 1835, and astronomers began looking for it well ahead of time; by then, of course, they knew more or less what to expect. Apparently the first observation was made on 6 August 1835 by Dumouchel, from Rome; the comet was seen as a dim, misty patch near the star Zeta Tauri, not far from the brilliant orange-red Aldebaran, the 'Eye of the Bull'. As the comet neared the Sun it brightened, and developed a major tail; it passed through perihelion in November, and though it faded quickly it was followed telescopically until May of the following year. The last observation of it was made by Sir John Herschel (son of Sir William) from his temporary observatory at the Cape of Good Hope, where he had gone to study the stars of the far south. Of course, this was in the pre-photographic era, but faithful drawings of the comet were made.

The latest return, so far, has been that of 1910. Very exact calculations of the orbit had been made by two Greenwich astronomers, A. C. D. Crommelin and P. Cowell, so that the comet was found early; it was detected on 12 September 1909 by Max Wolf in Germany, when it was still over 300,000,000 miles from the Sun—that is to say, well out in the asteroid zone, beyond Mars. (Let me add, however, that this applies only in the sense of distance. The orbital inclination of the comet is very high, so that it avoids the main asteroid swarm, and it moves in a retrograde or wrong-way direction, unlike any known asteroid.) It passed perihelion on schedule, within three days of the time given by Crommelin and Cowell, and it remained under observation until June 1911, when its distance from the Sun had grown to more than 500,000,000 miles—greater than the mean distance of Jupiter.

There were two points of special interest connected with this return. On 18–19 May Halley's Comet passed directly between the Earth and the Sun, and astronomers were anxious to know whether any sign of the event could be seen. The American astronomer Ferdinand Ellerman actually made a trip to Hawaii to observe under the best possible conditions, but he could see no trace of the comet in front of the Sun, which was extra confirmation of the flimsiness of cometary

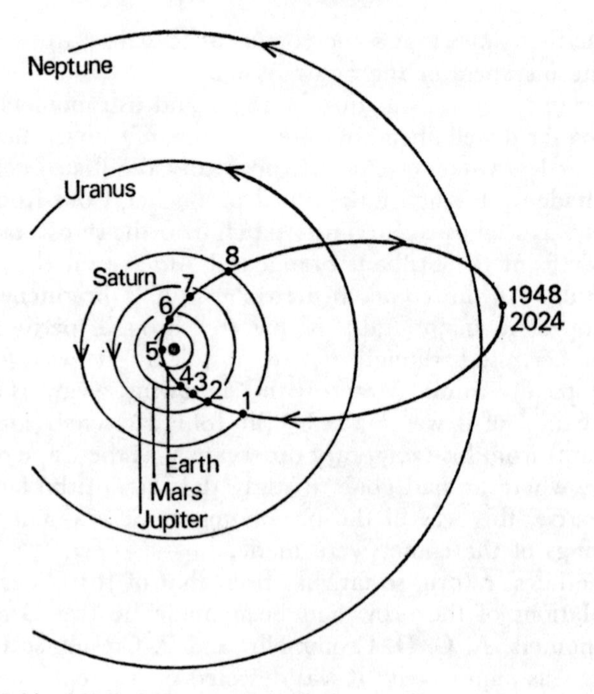


Fig. 14. Orbit of Halley's Comet. At perihelion the distance of the comet from the sun is less than that of Earth; at aphelion the comet recedes well beyond Neptune. The positions from 1983 to 1988 are shown. (1) mid-1983; (2) late winter 1985; (3) summer 1985; (4) winter 1985; (5) 5 February 1986; (6) spring 1986; (7) autumn 1986; (8) summer 1988. Perihelion will take place in 1986; the last aphelion was in 1948, and the next will be in 2024

material. Also, the Earth passed through the outer edge of the comet's tail, again without any visible result (though it is true that at this time the nucleus of the comet was still millions of miles away).

Over the years I have had many letters from people who tell me that they have vivid memories of Halley's Comet in 1910, and look forward to seeing it once more in 1986. Unfortunately, it seems that most of these accounts do not relate to Halley's Comet at all. Earlier in 1910 a much brighter, non-periodical comet had appeared; it became visible with the naked eye even with the Sun above the horizon, so that it was commonly known as the Daylight Comet. It was much more brilliant than Halley's, and I suspect that this is the object remembered by most of my correspondents, particularly as Halley's showed at its best as seen from the southern hemisphere. Actually, the Daylight Comet has been the most spectacular of the century so far.

At the 1910 return, Halley's Comet was picked up seven months before perihelion. The next return is due in February 1986 (preliminary calculations give the date of perihelion as 5 February), which means that it should be found again in the middle of 1985. It passed its aphelion, or furthest point from the Sun, in 1948—at a distance of some 3,300,000,000 miles, well beyond the path of Neptune. It is now drawing steadily closer to us, and it is speeding up, though we have no hope of finding it again yet awhile.

Just how brilliant Halley's Comet will become in 1986 remains to be seen. Quite possibly it is not so spectacular now as it used to be long ago, because, like all other periodical comets, it must suffer appreciable wastage of material every time it passes perihelion and develops a tail; but its 'life expectancy' must be long in comparison with a comet such as Encke's, which is smaller and returns more often. Yet we cannot be certain that there has been any measurable fading in historical times.

Records of it go back a very long way. The ancient Chinese were careful observers (even though they had no idea of what a comet was), and it is possible that a comet described by them in 467 B.C. was Halley's. The identification of the comet of 240 B.C. is much more definite. (Some catalogues use Old Style dates before the changeover in 1583, so that the date of the first identification of Halley's Comet becomes 239 B.C.) There are no reports of the return which presumably took place in 163 B.C., but the Chinese saw it again in 83 B.C. and also in 11 B.C. Since then it has been observed every time it has passed through perihelion: in A.D. 66, 141, 218, 295, 373, 451, 530, 607–608 (two bright comets were seen then, and we cannot be sure which was Halley's), 684, 760, 837, 912, 989, 1066, 1145, 1223, 1301, 1378, 1456 and then, as we have noted, 1531, 1607, 1682, 1759, 1835 and 1910. It has not always been brilliant; from all accounts it made a poor showing in 1145 and 1378, but was much more striking in 1301 and also in 1456, the year when Pope Calixtus III publicly excommunicated it as an agent of the Devil. The first known drawing of it refers to the return of 684, and comes from the *Nürnberg Chronicle*. It is certainly graphic, though I would hate to suggest that the representation is really faithful.

One celebrated return was that of 1066. The comet is said to have alarmed the Saxon Court, and there is a picture of it included in the Bayeux Tapestry, which some authorities believe to have been woven by William the Conqueror's wife.

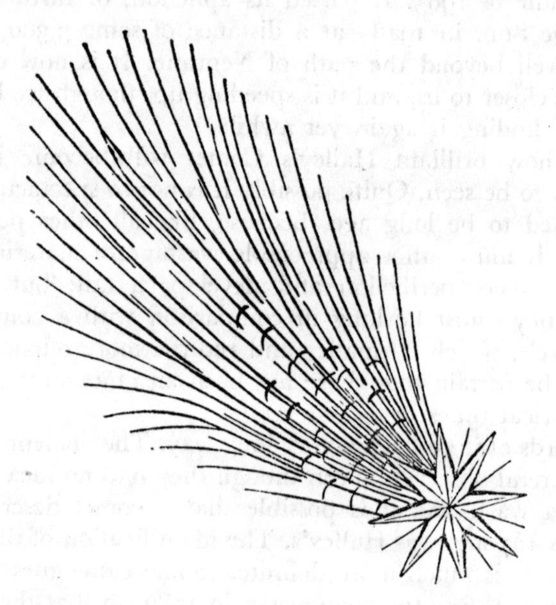

Fig. 15. Halley's Comet of A.D. 684 as drawn in the *Nürnberg Chronicle*

If the scene is authentic, King Harold is clearly toppling on his throne! A more scientific description of the comet in its 1066 guise has been given by a Greek historian named Zonares, who said that the comet looked at first as large as the Moon and had no tail; later, when the tail appeared, the comet was 'diminished in size'. Actually the comet must already have developed a tail when it became visible with the naked eye, and Zonares must have seen it head-on, so that the tail appeared as a luminous cloud around the comet's head.

Before leaving Halley's Comet, I think it worth while to

say something about a problem which has caused a great deal of interest in recent years. This concerns a possible tenth planet in the Solar System. There have been suggestions that the famous comet might provide us with an essential clue, though so far these hopes have not been borne out.

As we have noted, each planet pulls upon its fellows, producing what are called perturbations. The Earth's movement is affected by Venus, Mars, Jupiter and so on; if these planets did not exist, our orbit round the Sun would not be quite the same as it actually is. The perturbations can be worked out very accurately, and can be allowed for. It was in this way that the two outermost planets, Neptune and Pluto, were tracked down well before they were identified telescopically.

The story really began in 1781, when William Herschel, then an unknown amateur, was carrying out a 'review of the heavens' with a home-made reflecting telescope. In checking the stars in the constellation of Gemini (the Twins) he came across an object which looked quite unlike a star. It showed a definite disk, which no star can do, and it was in motion. Herschel believed it to be a comet, and indeed his original report to the Royal Society was headed 'An Account of a Comet'. Before long, however, it became clear that the object was a new giant planet, moving well beyond the orbit of Saturn, which had up to then been the most remote known body in the Solar System. The new planet was named Uranus, and Herschel's reputation was established.

Naturally, the mathematicians set to work, and calculated the way in which Uranus should move. Alas, the new world refused to behave. It persistently wandered from its predicted path, and eventually astronomers began to wonder whether it could be affected by yet another planet, further away from the Sun and as yet unknown. Careful calculations were made independently by Urbain Le Verrier in France and by John Couch Adams in England. The results agreed well, and in 1846 the predicted planet was found very close to the position which had been forecast.

With Neptune, as the newcomer was named, the Solar System again seemed to be complete, but there were still some very slight irregularities. In the early part of our own century the American astronomer Percival Lowell (best remembered today for his admittedly rather wild theories about the 'canals' on Mars), carried through a similar sort of investigation, and arrived at a position for a ninth planet. From his observatory at Flagstaff, in Arizona, he searched for it; but he failed to locate it, and after his death in 1916 the search was temporarily given up. Then, in 1930, a young astronomer at Flagstaff—Clyde Tombaugh, still, happily, very much alive—achieved success. He tracked down a very faint, starlike object which proved to be the long-awaited planet. It was named Pluto.

The reason for Lowell's initial failure was that Pluto turned out to be much fainter than expected. A telescope of some power is needed to show it, and it is a relatively small body, much less massive than the Earth. This set theorists a strange problem. A small planet, such as Pluto, could not produce any noticeable perturbations in the movements of giants such as Uranus and Neptune—and yet it was by these very perturbations that Pluto had been located. Something was wrong somewhere. Either the discovery had been sheer luck, or else Pluto must be much more massive than it seemed. Both these ideas seemed unlikely, and it was then suggested that Pluto might not be 'Lowell's planet' at all, in which case another world awaited discovery.

Certainly Pluto is an odd kind of world. Its orbital inclination is 17 degrees, much more than that of any other principal planet, and its path is so eccentric that it can come within the orbit of Neptune, as shown in Figure 16. For some years to either side of its next perihelion passage, due in 1989, it will forfeit its title of 'the outermost planet'. It is a slow mover; it takes over 247 years to complete one journey round the Sun, and it must be unbelievably lonely and dismal. All we know about its surface is that there seems to be a coating of methane ice.

Pluto is so much of an enigma that it may not even deserve to be ranked as a true planet at all. It could well be an ex-satellite of Neptune, which broke free in some unknown way and moved off independently. This idea is supported by the fact that Pluto is probably no larger than Triton, the senior of Neptune's two remaining satellites. If so, the possibility of there being a tenth planet seems very reasonable.

The main difficulty is that even if such a planet exists, it

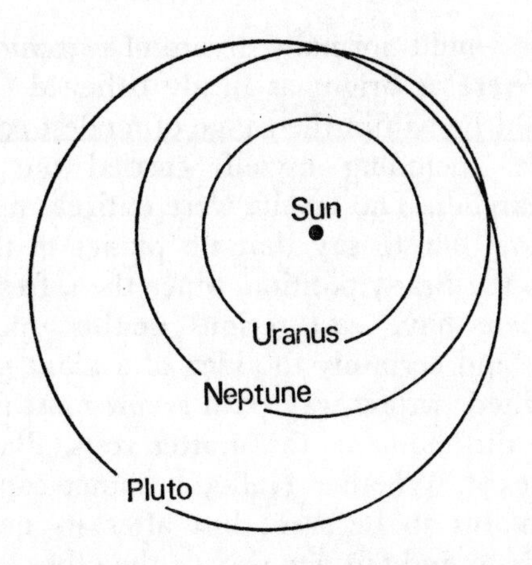


Fig. 16. The orbit of Pluto, which is relatively eccentric and inclined. At perihelion the distance from the sun is less than that of Neptune, but there is no fear of collision because Pluto's path is tilted at an angle of 17 degrees. The next perihelion is due in 1989

is bound to be extremely faint. Unless we have a really good idea of its position, finding it will be purely a matter of luck, and a systematic search would tie up a very large telescope for a very long time with no guarantee of success in the end. In 1972 Dr. Joseph Brady, of the United States, approached the problem from a different angle. Instead of considering the possible effects of Planet Ten upon the other planets, he concentrated upon the movements of Halley's Comet.

Remember, the orbit of the comet is very well known, because accurate measurements were obtained in 1835 and (particularly) in 1910. Working from these, Brady published a paper in which he discussed some minor discrepancies, and attributed them to a new giant planet with a mass three times that of Saturn, moving round the Sun in a retrograde orbit at more than twice the distance of Neptune. The revolution period was given as 512 years, and the inclination of the orbit was said to be as much as 60 degrees.

The position of Planet Ten, as given by Brady for early 1972, was in the northern constellation of Cassiopeia, whose five main stars make up the W or M shape familiar to anyone with even a rudimentary knowledge of the sky. Cassiopeia is a long way from the Zodiac, and would be the last place in which anyone would normally dream of searching for a major planet. If it were as bright as Brady believed (magnitude 14 to 15) it would be within the range of modest equipment, and many people—including myself—carried out prompt and systematic searches. The results were entirely negative, and I think it is now fair to say that no planet of this brightness exists close to the Brady position. Since then, fresh calculations made elsewhere have cast serious doubts upon the whole investigation, and certainly the idea of a giant planet moving in a highly-tilted, wrong-way path seems most improbable.

There, for the moment, the matter rests. Planet Ten may or may not exist. Whether Halley's Comet can provide any real clue remains to be seen, but after its next return we should be able to find out one way or the other.

Inevitably most of this chapter has dealt exclusively with Halley's Comet, but there are various others with periods of between sixty and 165 years which have orbits of the same general type. A few have been observed at more than one return, but they can never become brilliant, so that interest in them will be restricted to cometary enthusiasts and the average sky-watcher will not be concerned with them. But we do at least have the one bright, regular visitor, and everyone will be fascinated by Halley's Comet when it comes back once more in less than a decade from now.

*Chapter Seven*

# GREAT COMETS

THOUGH GREAT COMETS HAVE been so rare in our own century, they have been seen often enough in the past, and in this chapter I propose to say something about a few of them. Of course, earlier descriptions cannot be taken too literally, but some of them are straightforward enough; thus there is little doubt that the comet of 1264 had a tail which stretched more than half-way across the sky. On the other hand we also have accounts such as that of the comet of 1528, written by a French doctor named Ambroise Paré:

> This comet was so horrible, so frightful, and it produced such great terror that some died of fear and others fell sick. It appeared to be of extreme length, and was of the colour of blood. At the summit of it was seen the figure of a bent arm, holding in its hand a great sword as if about to strike. At the end of the point there were three stars. On both sides of the rays of this comet were seen a great number of axes, knives and blood-coloured swords, among which were a large number of hideous human faces, with beards and bristling hair.

One must suspect that Dr. Paré was less reliable astronomically than he may well have been in his own profession of medicine. There is even a doubt as to whether his description really relates to a comet at all, because the records do not show any spectacular visitor in 1528 (Halley's Comet, which came back in 1531, is certainly not involved). Quite possibly, what Dr. Paré saw was a brilliant display of aurora borealis (Northern Lights). Whether this is so or not, his description shows that comets were not welcomed.

At least there is no doubt about the comet of 1680, which was discovered by the German astronomer Gottfried Kirch, from Coburg. The tail was at one time 90 degrees long, and the nucleus was brilliant. The comet so alarmed the Rev. William Whiston that he predicted that it would eventually come back and destroy the world.

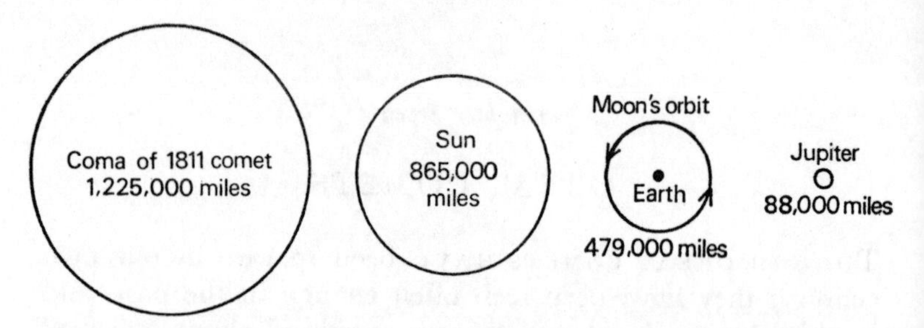


Fig. 17. The size of the coma of the Great Comet of 1811 compared with the diameters of the Sun and Jupiter and the orbit of the Moon. Despite its great size, the mass of the Comet of 1811 was negligible by planetary standards

The comet of 1744 was discovered on 9 December 1743 by Klinkenberg, in Holland, and was seen four days later by De Chéseaux in Switzerland; rather unfairly, it is always known as De Chéseaux' Comet. It must have been one of the most spectacular ever observed, because apparently it had at least half a dozen bright, broad tails. We have a good idea of what it looked like, because De Chéseaux himself left a drawing—made when the actual head of the comet was below the horizon, so that the tails swept upward in the manner of a fan. Unfortunately there are not many records of it, and it seems to have remained brilliant for only a few nights in early March 1744. Obviously we know nothing precise about its orbit, but it must have a period so long that it will not return for many centuries at least.

The Great Comet of 1811 has one distinction: it seems to have had the largest coma ever recorded (Fig. 17). At maximum the diameter was of the order of 1¼ million miles, so that it was considerably larger than the Sun even though its mass was so slight. Around October 1811 the tail extended for over a hundred million miles, with a breadth of fifteen million miles. It was under observation for some time; it was discovered by the French astronomer Honoré Flaugergues on 26 March 1811, and the last report of it came from Wisniewski, in Russia, on 17 August 1812.

The orbit was studied by F. W. A. Argelander, the German astronomer who was responsible for one of the most famous star-catalogues of the nineteenth century. Argelander gave the period of the comet as 3,065 years. He was very confident, and claimed that his estimate was correct to within fifty years either way. I confess that I am not so sure. Unfortunately I will not be able to check on the prediction in A.D. 4876, when the comet will presumably return if Argelander's calculations are right, but no doubt astronomers of that period will keep a careful watch!

As a casual aside, 1811 was also a year in which the port wine vintage in Portugal was unusually good. For years afterwards 'Comet Wine' was featured in the price-lists of wine-merchants, and advertisements of it lasted until 1880. Whether any 'Comet Wine' remains to be drunk today is, I suppose, problematical.

The only comet of near-modern times which surpassed that of 1811 was that of 1843. This seems definite, because a celebrated observer, Thomas Maclear, saw both, and wrote that the comet of 1811 'was not half so brilliant as the late one'. Moreover the tail was longer, and attained a record length of 200,000,000 miles (Fig. 18). According to descriptions

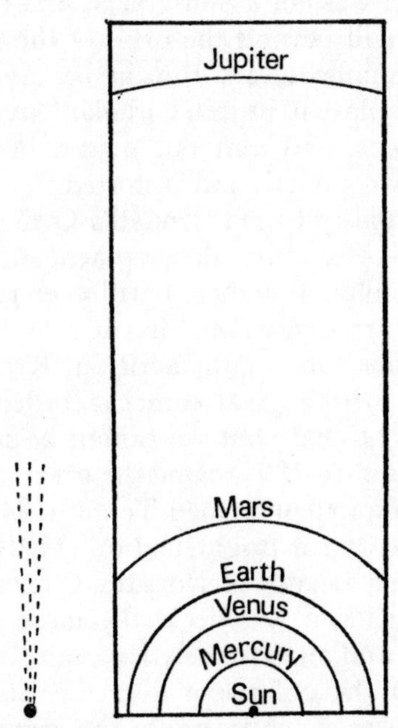


Fig. 18. Length of the tail of the Great Comet of 1843 compared with the orbits of the planets

and drawings of the time, the tail was relatively straight. The comet was of the type known as a 'Sun-grazer', and at perihelion it passed within about 100,000 miles of the brilliant solar surface, so that it must have been intensely heated. It was, of course, visible in daylight, and its orbit was practically or quite parabolic.

Next in our list of remarkable comets comes Donati's of 1858—possibly the most beautiful ever seen, because of its wonderfully curved main tail and its two shorter ones. Giambattista Donati, at Florence, discovered it on 2 June 1858; it was visible telescopically until 4 March 1859, and it remained a naked eye object for over three months. On 5 October 1858 it passed right in front of the brilliant star Arcturus—and Arcturus shone as brightly as usual; the flimsy material of the comet was quite unable to dim it. The tail grew from a length of 14,000,000 miles in late August to a maximum of over 50,000,000 miles in October, after which the length began to decrease.

Donati's Comet was not a Sun-grazer, and even at its closest approach it was still beyond the orbit of the planet Mercury. Its chief claim to fame, apart from its beauty, is that obvious disturbances took place in its tail. Circular 'envelopes' appeared outside the nucleus, and sent out masses of brilliant matter which passed down the tail and distorted it.

It is a tremendous pity that Donati's Comet made its entry—and its exit—before the development of reliable astronomical photography. Whether it will ever return we do not know, but there are suggestions that it may have a period of around 2,000 years. One mathematician, Kritzinger, believed it to be identical with a great comet recorded by the Roman writer Seneca in 146 B.C., but we cannot be sure.

Three years later, in 1861, came the great comet discovered by an Australian amateur named Tebbutt. It was then of the fourth magnitude, but it brightened quickly. It may not have had the surpassing beauty of Donati's Comet, but according to Sir John Herschel it was decidedly more brilliant, with a complex nucleus and an extremely long tail. In late June there is no doubt that the tail swept over the Earth—so that for some time we were actually inside the comet, though most people were blissfully unaware of the fact. The Earth did not pass through the coma, but the minimum distance of the head was only about 11,000,000 miles.

I have combed through papers and accounts written at the time in an effort to see whether anyone noticed anything unusual. The best testimony seems to have come from a meteorologist, E. J. Lowe, who claimed that the sky had a strange, yellow appearance on the evening of 30 June 1861, and that the brilliance of the Sun was reduced; in the parish church, one country vicar had the pulpit candles lighted at 7 o'clock. Lowe also mentioned that the comet looked hazier than on any other evening. So far as I know, this comet was the first to be the subject of attention from an astronomical photographer. One of the great English pioneers, Warren de la Rue, tried to record it with his primitive equipment, though without success.

I must pass quickly over the other bright comets of the later nineteenth century. The Great Comet of 1862, discovered by Lewis Swift, did not equal that of the previous year, but it was nevertheless striking, and was notable for the luminous jet which came from its nucleus. Coggia's Comet of 1874 was also brilliant, and so was that of 1880, best seen from the southern hemisphere.

The comet of 1882 rivalled any of its predecessors, and was excellently photographed by Sir David Gill from the Cape of Good Hope—the first good comet picture ever obtained. At one stage, according to several observers (including Barnard and Brooks) the comet threw off a mass of luminous material which gave the temporary impression of making up a satellite comet.

There was an important sequel to Gill's work. When he photographed the comet, he also recorded many stars, and it was this which made him realize the potential value of stellar photography. This led on to the compilation of the photographic star atlases which have now completely replaced the older methods of star-cataloguing.

There was another fairly brilliant comet in 1887, and yet another in 1901, discovered by Paysandu in South America and which was best seen from the southern hemisphere; apparently it was distinctly yellowish in colour. Then came the Daylight Comet, discovered by some diamond miners in the Transvaal on 12 January 1910. Needless to say, they were not looking for comets, but they could hardly overlook this one. By 21 January it was well visible from England, and I quote from a description given by E. Hawkes, observing from Leeds:

> The comet was picked up with the naked eye at 4h 40m, and was a gorgeous object. The picture presented in the western sky was one which will never be forgotten. A beautiful sunset had just taken place, and a long, low-lying strip of purple cloud stood out in bold relief against the glorious primrose of the sky behind. Away to the right the horizon was topped by a perfectly cloudless sky of turquoise blue, which seemed to possess an unearthly light like that of the aurora borealis. High up in the south-west shone the planet Venus, resplendently brilliant, while below, and somewhat to the right, was the great comet itself, shining with a fiery golden light, its great tail stretching some seven or eight degrees above it. The tail was beautifully curved like a scimitar, and dwindled away into tenuity so that one could not see exactly where it ended. The nucleus was very bright, and seemed to vary. One minute it would be as bright as Mars in opposition, while at another it was estimated to be four times as bright. The tail, too, seemed to pulsate rapidly from the finest veil possible, to a sheaf of fiery mist.

No doubt these apparent fluctuations were due to conditions in the Earth's atmosphere, but the description is graphic, and it agrees excellently with the account given to me years ago by my colleague the late H. Percy Wilkins, who saw the comet from Wales at about the same time. There is no doubt that it was far superior to Halley's, which appeared later in the same year; but Halley's, of course, is a regular visitor, whereas the Daylight Comet will not be seen again in our time.

After 1910 there was a long hiatus, though Skjellerup's Comet of 1927 was bright for a few nights as seen from the southern hemisphere. Other brief visitors were the comets of 1947 and 1948, both of which were bright and had long tails. The 1948 comet was discovered during a total eclipse, seen from Africa, but by the time it came into view in the northern hemisphere it was well past its best.

There have, of course, been naked-eye comets of lesser splendour. I well remember Finsler's of 1937, and Comet

Jurlov-Achmarov-Hassell of 1939, which had a pronounced greenish hue. But northern observers had to wait until 1957, when there came two comets, both non-periodical. By no stretch of the imagination could they be classed as 'great', but they were bright enough to cause general interest even among non-astronomers.

The first of them was discovered by two Belgian astronomers, Arend and Roland. It was easy to see with the naked eye, and when I looked at it with binoculars, in late April, its tail was splendidly displayed. Perhaps I should say 'tails', because ahead of the comet there could be seen a remarkable spike. Actually, this was due to thinly-spread material lying in the plane of the comet's orbit; the material was much too diffuse to be easily visible far from the head when observed broadside-on, but when the layer was seen 'from the edge' it gave the false impression of a sunward tail. Arend-Roland had a nucleus which reached the first magnitude, and it remained in telescopic view for some months. In August 1957 another fairly bright comet made its entry; it was discovered by the Czech astronomer Antonín Mrkós, and it too became conspicuous with the naked eye, though it lacked any spike and remained prominent for only a relatively brief period in the morning sky.

It seemed reasonable to hope that the barren years were over, and that new bright comets would appear. Pereyra's of 1963 would have qualified as 'great' but for the fact that it was never well placed, and came nowhere near the Earth, so that its full glory was completely lost. Seki-Lines, in 1962, made a brief showing. Then came Ikeya-Seki of 1965, discovered by two of the energetic Japanese searchers on the morning of 21 October. Amateur and professional observers all over the world were very much on the alert, since one magnitude estimate had been given as $-9$, in which case the comet would have been much the most brilliant since the far-off days of 1910. Undoubtedly it was a Sun-grazer, and there had even been a suggestion from the Moscow astronomer B. Y. Levin that it would actually hit the Sun.

Needless to say, this announcement sparked off all kinds of rumours. One 'expert' claimed that the results of such a collision would be to disrupt television reception; the editors of a London daily paper published a spectacular picture of what they fondly believed to be the comet, but was in fact an aircraft condensation trail. Subsequently Levin changed his mind, and announced that the comet would miss the Sun by a small margin, though the intense heat might disrupt it.

At this time I was Director of the Armagh Planetarium, in Northern Ireland. Since the sky was cloudy, I prevailed upon the Royal Air Force to take me up in a high-flying aircraft. From a vantage point above the clouds I hoped to obtain good photographs, but the comet obstinately refused to show itself. Dr. D. W. Dewhirst, of Cambridge University, was similarly unsuccessful from an aircraft over Southern England, even though the comet was then scheduled to be at its best as it emerged from behind the Sun. Naturally we were disappointed, and it transpired that once again northern-hemisphere observers had been unlucky. Toward the end of October and the beginning of November magnificent views were obtained from South Africa, South America, Australia and parts of the United States, and there is no doubt that the comet was a 'great' one, with a brilliant head and a long, slightly-curved tail extending to over 30 degrees. It was even said by some observers to be comparable with the Great Comet of 1882. Considerable activity took place in the nucleus around the time of perihelion, and when it emerged from the Sun's rays it was described as having a distinctly yellowish hue. At its minimum distance from the Sun it was a mere 307,000 miles from the solar surface, and spectroscopic examination showed the presence of atoms of sodium, iron, nickel, copper, calcium and other elements—but the familiar signs of molecules were lacking, since the molecules, which are groups of atoms, had been broken up by the intense heat.

Another Japanese-found comet, Tago-Sato-Kosaka of early 1970, was rather disappointing as a spectacle, but was scientifically important because of the discovery that it was enveloped in a huge cloud of rarefied hydrogen. It was followed in the spring by something really worth seeing—the comet which had been discovered late in the previous year by Jack Bennett, from Pretoria. When at its best, Bennett's Comet was well south of the equator of the sky, but it was still bright when it came northward, and I had some splendid views of it from my Selsey observatory throughout April. Both types of tail were displayed; the 19-degree-long curved tail made of dust, and the shorter, straight tail composed of gas. Spiral-shaped jets were seen in the nucleus, and the whole comet showed marked activity. It, too, was enveloped in a hydrogen cloud, and the diameter of this cloud seems to have been about 8,000,000 miles, though of course the material was not visible telescopically.

In early April I estimated the magnitude of the nucleus as 1·2, slightly brighter than the star Deneb in Cygnus. Even then it was fading, and by now it has long since disappeared altogether; it has started its long journey back into the depths of space, and we will not see it again.

We now come to a comet which promised to be 'great', but failed to come up to expectations. This was Kohoutek's Comet,* which caused a tremendous amount of general excitement as it approached the Sun and the Earth during 1973.

It was discovered on 7 March 1973 by Dr. Lubos Kohoutek, a Czechoslovak astronomer who works at the Hamburg Observatory in Germany. The circumstances were distinctly unusual. Dr. Kohoutek was looking for a faint periodical comet, Biela's, which had not been seen since 1852, and which has certainly broken up. When he developed his plates, he found a comet which was definitely not Biela's, and which he soon realized was something unusual. It was faint, with clear indications of a coma. The exceptional thing about it was its distance: over 400,000,000 miles from the Earth and over 430,000,000 miles from the Sun. Few comets are visible as far out as this, so that the newcomer was obviously very large.

Further calculations made the outlook seem even more promising. During much of the summer the comet would be on the far side of the Sun, but it would emerge into the morning sky during October, and would cross the Earth's orbit in November, though at that time the Earth would be some distance away and there was not the slightest chance of

* Not to be confused with Kohoutek's periodical comet, which was found by the same observer with the same telescope, but which is always faint. It is customary to refer to periodical comets with the symbol P/; thus Halley is P/Halley, Encke is P/Encke, Tempel 1 is P/Tempel 1, Kohoutek's periodical comet is P/Kohoutek and so on. Dr. Kohoutek was also one of the discoverers of P/West-Kohoutek-Ikemura, in 1975.

a collision. Predictions were that it ought to become a naked-eye object in November, and that at its brightest it would attain a magnitude of something like −12, comparable with the full moon. This would naturally make it a daylight object, comparable with any of the comets of past years. At perihelion, on 28 November 1973, it would be racing along at about 100 miles per second (as against the modest 18½ miles per second, or 66,000 m.p.h., for the Earth) and it would pass within 13,000,000 miles of the Sun itself. After perihelion, when it had come into the evening sky, it would still be magnificent, and would by-pass the Earth at about 75,000,000 miles in mid-January 1974. It was not expected to fade below naked-eye visibility until the end of February, and there was every reason to suppose that it would develop a really long tail. Writers went so far as to refer to it as 'the Comet of the Century'.

Astrologers and their kind were in full cry at an early stage, and all sorts of dire predictions were made. The religious tract put out by an organization calling itself the Children of God, and written by a Mr. Moses David, referred to the announcement of the comet as 'shocking', and claimed that the maximum brightness would be seven times greater than that of the Moon. In this tract, headed *The Christmas Monster*, Mr. David went so far as to ask whether the apparition heralded the end of 'Fascist America and its new Nazi Emperor', and a few remaining end-of-the-worlders also voiced their apprehensions. Scientists approached the matter differently. It was thought possible that Kohoutek's Comet came from a 'comet cloud' orbiting the Sun at a distance of about one light-year (approximately 5,880,000,000,000 miles) and was paying its first visit to the Sun, after a journey lasting for two million years. If so, then it would be very 'dusty', since it would never before have suffered evaporation of the material in the nucleus under the influence of solar heat.

But as October passed by, there were uneasy doubts as to whether Kohoutek would prove brilliant after all. The magnitude was well below the predicted value, and the tail was negligible. Though low in the dawn sky, it lay in the right position; but from my Selsey observatory I failed to see it at all until 17 November, when I estimated the magnitude as 7. I could detect only a trace of a tail, even though I was using my 15-inch reflector—which, by amateur standards, is powerful. By the end of November the magnitude had risen to 6, and there was a quarter-degree tail. Observers in clearer climates were able to follow it through most of December, when the tail grew to more than 10 degrees in length even though it was not brilliant. So far as I know, Jack Bennett, from Pretoria, was the last to see it before its perihelion passage on 28 December; his final view was on the 22nd, when it was very low in the dawn glare. There was certainly no chance of its being glimpsed in daylight, as the comets of 1811, 1843, 1861, 1910 and others had been.

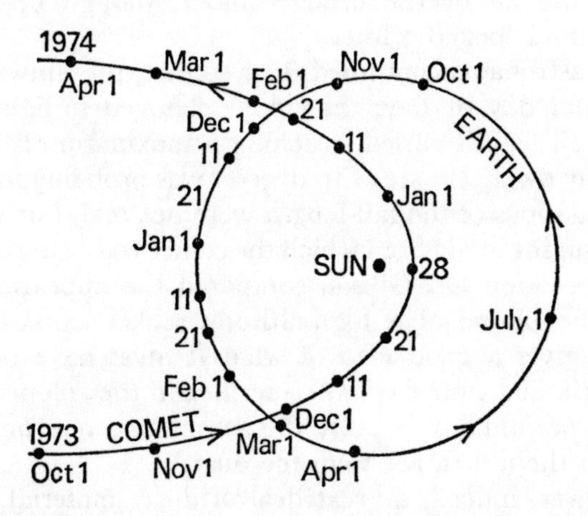


Fig. 19. Orbit of Kohoutek's Comet and positions of the earth, October 1973 to April 1974

During early January 1974 it reappeared in the evening sky, but it was still disappointing, and by now members of the general public were becoming sadly disillusioned.* I had my best view of it on 10 January—not from my home, but from an aircraft flying at 33,000 feet above the Irish Sea. True, the brilliant planets Venus and Jupiter, together with the comet, made up a fascinating spectacle; but the magnitude of the comet was below 3, and the tail cannot have been more than 10 degrees long according to my estimate. By the end of January the comet had fallen well below naked-eye visibility,

* In an article in the London *Times*, my old friend Bernard Levin asked plaintively what had happened to the comet, and accused me of having stolen it!

and it continued to decline, though it was followed telescopically for many months.

Yet the view from above the Earth's atmosphere was very different, and fortunately this was the time when the United States vehicle Skylab was manned by its last crew—Astronauts Carr, Gibson and Pogue. During a 'space-walk' outside their craft, only a day after the comet passed perihelion, they had an excellent view from their vantage point some 270 miles above the ground. There was a distinct tail, and a sunward 'spike' recalling Arend-Roland of 1957. Dr. Gibson described it as 'awe-inspiring', with an exceptionally bright nucleus, and an overall orange colour, though before perihelion it had looked white.

As the astronauts continued their studies, the sunward spike grew fainter day by day; the colour changed to light yellow, and the tail length varied, reaching a maximum of 8 degrees (so that my own estimate of 10 degrees was probably too great). The fluctuations of the tail-length were not real, but were due to the changing angle from which the comet was being observed. On one occasion Dr. Gibson compared the appearance with that of the plume of a high-altitude rocket exhaust, which certainly gives a good idea of what it must have been like. Between six and nine days after perihelion the colour changed to violet, presumably because the astronauts were now seeing the gas in the tail rather than the dust.

There was, indeed, a great deal of dusty material, as both Earth-based and space observations showed, but there was one more major disappointment. Researchers at the United States National Radio Astronomy Observatory, at Green Bank in West Virginia, used the 36-foot radio telescope to search for various molecules in the cometary material, but unfortunately the position which they had been given was in error by 45 seconds of arc, so that while they were carrying out their measurements, between 14 and 18 January, they were pointing their equipment in the wrong direction. It is hardly surprising that the results were negative. One might even say that Kohoutek's Comet was awkward to the last!

At least the comet's movements were carefully and accurately studied. The orbit proved to be inclined at an angle of 14·3 degrees, and to be highly elliptical. The estimated period is

75,000 years, and if this is approximately correct the aphelion distance is of the order of 320,000,000,000 miles, roughly ninety times the mean distance of Pluto.

Finally, we come to West's Comet of 1976, which was one of the finest of the century and must be at least on the verge of being termed 'great'. It was first identified by Richard West, a Danish astronomer stationed at the Geneva headquarters of the European Southern Observatory (E.S.O.). Photographic work for the E.S.O. is carried out at La Silla, in Chile; and when examining plates which had been taken in August and September, West discovered the image of a faint comet, no brighter than the sixteenth magnitude. The first plate to show it had been taken on 10 August 1975, but West's announcement was not made until 5 November.

As the comet drew inward it brightened up. Whereas Kohoutek's Comet had always been below its predicted brilliancy, West's was brighter by about two magnitudes, and even before it reached perihelion it had become spectacular. When at its closest to the Sun it was still some 18,300,000 miles from the solar surface, so that it did not qualify as a Sun-grazer, but already it had developed tails of both gas and dust. Only seventeen hours after perihelion, on 25 February 1976, the well-known observer John Bortle, in New York State, saw it with the naked eye in full daylight, only 7 degrees from the Sun; he estimated the magnitude as −3, brighter than Jupiter or any other planet apart from Venus.

During March, West's Comet made a fine display in the morning sky. The gas tail was long, slender and straight; it was distinctly bluish in colour, whereas the broad, curved and fan-like dust tail was white. The dust tail showed delicate, fibrous structure, well brought out in the countless pictures taken; it has been claimed that West's was the most-photographed comet of all time. Inside the head was a bright, compact nucleus, appearing starlike with the naked eye and about a magnitude fainter than the total light of the coma.*

During the latter part of March the comet faded, the magnitude had dropped to about 4 by the 22nd of the month,

* Unfortunately I have to rely mainly on second-hand reports. When the comet was at its most brilliant I was confined to bed with influenza—a typical case of Spode's Law: if things *can* go wrong, they *do*.

still about two magnitudes brighter than had been originally predicted. The tail also shrank, and as it receded from the sun the comet showed obvious signs of breaking-up. It had suffered badly during its headlong dash past perihelion, and the damage to it was irreparable. When I last saw it, on 25 May, my 15-inch reflector showed three very dim, hazy condensations; it was hard to believe that the comet had been so glorious only a couple of months earlier.

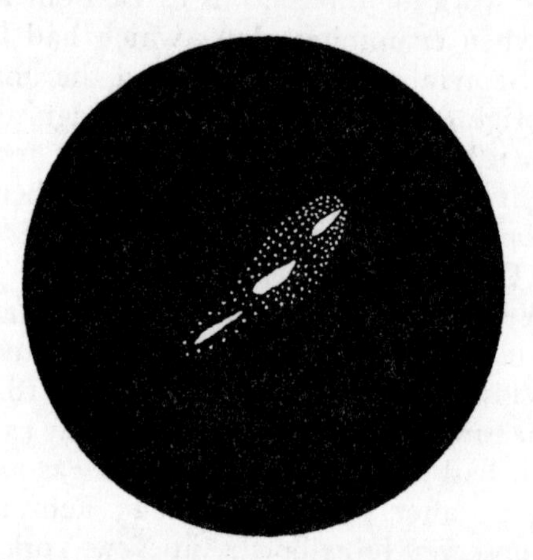

Fig. 20. Sketch of West's Comet made on 25 May 1976 at 23.00 hours by Patrick Moore using a 15-inch reflector × 200. The comet has already started to break up and has become very faint

According to the best available calculations, the period is about 300,000 years. Yet even if West's Comet survives to return after so long, it can never again become as brilliant as it was in 1976. There is no chance of its providing a repeat performance.

So much, then, for West's Comet, now back in the depths of space from whence it came. When we will see a more imposing visitor, we cannot tell. We can only hope that there will be more truly great comets in the near future than there have been in the recent past.

*Chapter Eight*

# LOST COMETS

MOST OF THE CELESTIAL bodies have very long lives. The Earth and its Moon have existed for at least 4,700 million years, and the Sun is presumably even older; the planets move in stable orbits, and there is no reason to expect that any disaster will overtake them. Not so with the comets, which are fragile and ephemeral. They die quickly on the cosmical time-scale, and we have even seen one or two of them in their death-throes.

Of course, a non-periodical comet (by which I mean a comet moving in either a parabola, hyperbola or very eccentric ellipse) will be lost as soon as it moves out of the inner part of the Solar System, and will not be recovered; thus we can no longer track Donati, Arend-Roland, Bennett, Kohoutek and many others, but there are no grounds for suggesting that they have broken up. Comets of shorter period may be lost either because their movements are not sufficiently well known, or because of genuine disintegration.

If a short-period comet is not adequately observed over a period of at least several weeks, its orbit is bound to be uncertain, and it may not be found again at subsequent returns. This has happened many times, as is only to be expected. We also have comets which are so violently perturbed by Jupiter (or, less frequently, by some other planet) that we are unable to locate them afterwards. This brings me on to the classic case of Lexell's Comet.

The comet was discovered by Messier on 14 June 1770. It was visible with the naked eye, and its apparent diameter was given as five times greater than that of the full moon—not because it was exceptionally large, but because it was exceptionally near; it passed within 750,000 miles of the Earth. The orbit was calculated by the St. Petersburg mathematician Anders Lexell, who allotted a period of 5½ years. Unfortunately Lexell did not publish his results until 1778, by which time the whole situation had changed. He found that before 1767 the period had been much longer, but then an encounter with Jupiter had thrown the comet into the orbit which brought it close to the Earth. Another approach, in 1779, produced a further alteration, the net result being that the comet kept too far away from us to be seen at all. It has never been located since, and we can hardly hope to recover it now.

I must, I feel, mention a remarkable theory which was given in a book published in 1811—written originally by James Ferguson, who began his career as a shepherd-boy in Scotland and ended it as a famous writer on astronomy. Ferguson was born in 1710, and died in 1774. After his death, his book was revised by Sir David Brewster, who referred to the four minor planets or asteroids which had been discovered between 1801 and 1808, and went on as follows:

> It is a very singular circumstance, that while two of the fragments, Juno and Vesta, are entirely free from any nebulous appearance, the other two fragments, Ceres and Pallas, are surrounded with a nebulosity of a most remarkable size. In the case of Ceres, this nebulosity is 675 English miles high; while the nebulosity of Pallas extends 468 miles from the body of the planet. It is obvious, that such immense atmospheres could not have been derived from the original planet, otherwise Juno and Vesta would also have been encircled with them; so that they must have been communicated to Ceres and Pallas, since the planet was burst.

Brewster was assuming, as some people still do today, that the asteroids were the débris of an old planet which disintegrated in the remote past.

> Now, the Comet of 1770, if it is lost, must have been attracted by one of the planets whose orbit it crossed, and must have imparted to it its nebulous mass; but none of the old planets have received any addition to their atmosphere; consequently, it is highly probable that the Comet has passed near Ceres and Pallas, and imparted to them those immense atmospheres which distinguish them from all the other planets.

Fascinating indeed—but, of course, completely wrong. No planet could be endowed with an atmosphere in such a way, and in any case neither Ceres nor Pallas has any vestige of atmospheric surround. Their escape velocities are far too low, and their images are as sharp and clear-cut as those of all the other asteroids.

Another interesting comment was made by the great French mathematician Pierre Simon de Laplace (of Nebular Hypothesis fame), who pointed out that if Lexell's Comet had been as massive as the Earth, its effect during the encounter of 1770 would have been to shorten the length of our day by 2 hours 47 minutes. The fact that nothing of the kind was observed made Laplace conclude that the mass of the comet was no more than 1/5000 of that of the Earth. Actually, the true value is very much less than this.

Lexell's Comet has vanished from our ken simply because it is now very faint, and we do not know where to look for it; it has not disintegrated. But with other comets, the causes of 'loss' are much more fundamental.

Consider Westphal's Comet, first detected in 1852 and at that time described as 'pretty bright'. It belonged to the so-called long-period class, with a revolution period of sixty-one years. It came back on schedule in 1913, but as it drew in toward the Sun it faded out, and evidently did not survive its perihelion passage. There was no question of its having been pulled out of position into an unexpected orbit, and it must be regarded as defunct. The next return to be predicted was that of January 1976, but no trace of it was seen. Brorsen's periodical comet, first detected in 1846, is another such casualty; and in 1926 Ensor's Comet became diffuse as it approached perihelion, and it too vanished as effectively as the hunter of the Snark.

Yet one must beware of jumping to conclusions. In 1892 E. E. Holmes discovered a comet with a period of approximately seven years; it was dimly visible with the naked eye, and was therefore one of the brightest of the short-period comets. At its subsequent returns it was fainter, and after 1906 it 'went missing'. It was listed as being as dead as any dodo, but one energetic astronomer, Dr. Brian Marsden, was not satisfied. He made an exhaustive series of calculations, and as a result Holmes' Comet was found again at its return in 1964—though it had become very faint indeed. Marsden was equally successful in tracking down another comet, Di Vico-Swift, which had not been seen since 1894 even though its period was a mere 6·3 years. It was finally found again in 1965.

The latter comet has an interesting history. In 1844, when Di Vico discovered it, it was on the fringe of naked-eye visibility, and had a short tail. It was badly placed at the next return, that of 1850, and nobody was surprised when it was missed. Yet it should have been well seen in 1855, and when it could not be located it was regarded as lost.

Then, in November 1894, Swift in California discovered a telescopic comet whose orbit was so like that of the lost Di Vico that there could be no serious doubt about the identity. Jupiter had been the culprit once more; a close approach to the giant planet in 1855 had twisted the orbit into a period of only 5¾ years. Yet another encounter, in 1897, changed the period again, this time to 6·4 years; and it was understandable that the comet was lost once more.

Nothing further transpired until the 1960s, when Marsden took up the problem. His calculations proved to be amazingly accurate, and under the clear skies of Argentina Dr. Arnold Klemola, in 1965, managed to photograph the comet just where Marsden had said it would be. (I suggest that it should be termed Marsden's Comet rather than the official designation of Di Vico-Swift.) The discovery was timely, because in 1968 a third encounter with Jupiter threw the comet back once more into a larger orbit. This means that it cannot now become bright enough to be seen with the naked eye, or even with binoculars.

Various other comets have been found, had their orbits calculated, been put into the short-period catalogues and have then 'gone missing', generally because their paths have not been observed for sufficient lengths of time. W. F. Denning, one of Britain's most famous observers of comets and meteors, found two comets, one in 1881 and the other in 1894, which should have had periods of 8·7 and 8·2 years respectively; neither was ever recovered. Other cases are those of Comet Barnard 1 of 1884 (5·4 years according to calculations), Brooks of 1886 (5·6 years), Spitaler of 1890 (6·4 years), Barnard 3 of 1892 (6·5 years—incidentally, the very first periodical comet ever to be discovered by means of photography), Swift of 1895 (7·2 years), Metcalf of 1906 (7·8 years) and Schorr of 1918 (7 years). But there was also Taylor's

Comet of 1916, whose period was worked out as 6·4 years. As it receded after its first appearance, it divided into two parts, and was given up for lost—presumably because of disintegration. It came as a major surprise when, in December 1976, C. T. Kowal, using the Schmidt telescope at Palomar, discovered a comet which corresponds so exactly to Taylor's that we are forced to the conclusion that it *is* Taylor's. It had been lost for sixty-one years, and had made nine circuits of the Sun since it had been originally found.

We must, then, take all due precautions before mourning the death of a comet. But in the case of the most famous lost comet of all, Biela's, there can be no doubt whatsoever, and this brings me on to another very important topic—the close association between comets and meteor streams.

In 1772 Charles Messier discovered a faint comet which seemed to be entirely unremarkable. It was also seen by another French observer, Montaigne, and was kept under observation for some time. In 1810 Friedrich Bessel—the German astronomer later to become famous as being the first man to measure the distance of a star—checked on the orbit of the 1772 comet, and decided that it must be identical with a comet which had been seen in 1806. If it had a period of 6¾ years, it had returned several times during the interim without being detected, which was not in the least surprising.

Bessel predicted a return for 1826, and alerted his colleagues. One of these was an Austrian soldier, Captain Wilhelm von Biela, who was a skilful amateur astronomer. Von Biela managed to capture the comet; he saw it on 27 February 1826, and worked out that its period must be 6 years 9 months. Ten days later it was also seen by the French astronomer Adolphe Gambart, and it is sometimes—unjustifiably, I feel—referred to as Gambart's Comet.

The next return, that of November 1832, was entirely normal. The perihelion passage of 1839 was missed, because the comet stayed too close to the Sun in the sky, but there seemed no reason to doubt that Biela's Comet would be back once more in 1846, and in late 1845 it duly made its entry. To everyone's surprise, it was not alone. It was accompanied by a second luminous patch, which developed into a second comet. The original object had split in two.

A detailed description was given in late January 1846 by Professor James Challis, at Cambridge University, who was using the 12-inch Northumberland refractor. (Challis is best remembered today for not having discovered the planet Neptune, even though all the information had been put into his hands; but that is another story.) Challis said:

> There are certainly two comets. The north preceding is less bright and of less apparent diameter than the other, and has a minute stellar nucleus . . . I think it can scarcely be doubted, from the above observations, that the two comets are not only apparently but really very near each other, and that they are physically connected. When I first saw the smaller, on 15 January, it was faint, and might easily have been overlooked. Now it is a very conspicuous object, and a telescope of moderate power will readily exhibit the most singular celestial phenomenon that has occurred for many years—a double comet.

Astronomers were intrigued, and waited eagerly for the next return, that of 1852. This time the two comets were separated by over a million miles, and remained on view for several weeks.

In 1859 the position of the comet(s) was again so unfavourable that no observations could be expected, but conditions in 1866 should have been ideal, and searches began well ahead of time. They proved to be completely fruitless. Though the predicted position of the comet pair had been worked out with great precision, absolutely nothing was seen, and Biela's Comet was conspicuous only by its absence.

Apparently nothing could be done, and nobody had any real hope of finding the comet again at the next predicted return, that of 1872. Yet there was another line of investigation. There had long been a suspicion that meteors might be related to comets, and that meteoritic material could well be spread out along the orbits of known comets; Giovanni Schiaparelli, best remembered now for his observations of the 'canali' on Mars, had already found that the brilliant Perseid meteors moved in the same path as that of a periodical comet (Swift-Tuttle). Two German astronomers, E. Weiss and H. L. D'Arrest, calculated that a meteor shower observable in late November each year could be associated with Biela's Comet, and that there might be a major display in 1872, at the time when the comet ought to have been on view. Weiss gave the probable date as 28 November, D'Arrest as 9 December.

Weiss proved to be more or less correct. On 27 November a brilliant meteor shower was seen, and there can be no doubt that these meteors represented the débris of the dead comet. Since then the 'Bieliid' shower has become much weaker, but a few meteors are still seen each year to come from the old source, so that even today we have not lost all trace of Biela's Comet.

Why did the comet vanish so completely? There have been suggestions that the original division into two parts was caused by a close approach to Jupiter in 1842, and that the pull of the Sun did the rest. But there is another episode which deserves to be put on record, even though it has never been fully explained. One astronomer who kept up his interest in the comet even after the fiasco of 1865 was Wilhelm Klinkerfues of Göttingen, and on 30 November 1872 he sent a cryptic message to his colleague Pogson, at Madras in India: 'Biela touched Earth on 27th; search near Theta Centauri.' (Theta Centauri is a bright southern-hemisphere star, never well seen from the United States and invisible from Britain.) Pogson made a search, and on 2 and 3 December 1872 he actually observed a comet. Alas, bad weather and the approach of twilight conditions prevented him from seeing it again, and nothing more is known about it; but Pogson was a highly experienced observer, and he described his comet as having a bright nucleus and an appreciable tail, so that there seems little room for error. On the other hand, it is inconceivable that the comet was Biela's; it must have been another, quite unconnected, merely happening to lie in the same region of the sky—an almost incredible coincidence. After the lapse of more than a century, we can scarcely hope to solve the mystery now.*

* Another double comet was that of 1860, seen by the French astronomer Liais from Olinda in Brazil, and always known as the Olinda comet. Liais saw the two comets on 27 February, the day after the discovery; he followed the pair until March, but on 13 March he reported that the secondary member had disappeared. Unfortunately nobody else saw the Olinda comet at all, so that little more can be said about it, and some astronomers tend to doubt the authenticity of Liais' observations.

Other comet-meteor associations are equally beyond doubt. For instance there is the meteor shower connected with a famous periodical comet, Giacobini-Zinner. The comet itself is not remarkable in appearance; it never becomes visible with the naked eye, though its orbit is well known, and there are even suggestions that a rocket probe may be sent to it during the 1980s. It was discovered by Giacobini, from Italy, in 1900; it was missed at its 1906 return, but was recovered in 1913 by Zinner (hence the double name), and has been seen regularly since then. In 1933 the Earth made contact with the meteor swarm moving in the comet's orbit, and for an hour or two there was a positive rain of meteors. Over some parts of Europe, more than 15,000 meteors were estimated to have been recorded within sixty minutes. A second shower of comparable intensity was seen in 1946, and this time astronomers were ready for it, so that it was well studied. Unfortunately, the swarm was subsequently perturbed by Jupiter, and we cannot be sure when or if we will again see a major display of 'Giacobinids'.

Various other meteor showers are known to be connected with comets: the Beta Taurids (24 June to 6 July yearly) with Encke's Comet; the other Taurids (15 September to 15 December, with a maximum around 14 November) also with Encke's; and both the Eta Aquariids (21 April to 12 May) and the Orionids (18 to 26 October) with Halley's, though it is true that this latter association has been questioned. Then, of course, we have the Leonids of 17 November, which are linked with the orbit of a faint periodical comet, Tempel 1, first seen in 1866. The Leonids are not reliable. They used to produce major displays approximately every thirty-three years —in 1799, 1833 and 1866—but planetary perturbations robbed us of the promised displays of 1899 and 1933. In 1966 the conditions seemed favourable, and observers waited eagerly. In England, I devoted a television programme in my *Sky at Night* series to the shower, and asked enthusiasts to watch between midnight and dawn to record any meteors which might appear. We were disappointed. Very few Leonids were seen, and one disgruntled viewer wrote to me: 'Watched from midnight until 6 a.m. on 17 November. Meteors: from the sky—none. From the wife—plenty.' In fact, the display occurred during daylight over Europe, so that we missed it by a few hours. From other parts of the world, such as Arizona, it was magnificent, with a rate of more than 100,000 shooting-stars per hour. The fact that the shower lasted for so short a time proved that the meteors are 'bunched' rather than being spread more uniformly along the comet's orbit, as happens with the Perseids; ironically, our observations from the British Isles were more valuable scientifically than they would have been if we had seen the main display. Whether we will be treated to a further shower of celestial fireworks in 1999 depends upon whether the orbit is again shifted in the interim.

It would be an over-simplification to say that meteors are due to the breaking-up of old comets, but of the close association there can be no doubt whatsoever. Remember, the solid particles in a comet are of meteor size, which is why the 'dust' in the tails can be repelled by the stream of low-energy particles coming from the Sun.

What, then, about meteorites? Here there is no evidence of any cometary association, and it is certain that meteorites are much more nearly related to asteroids than to shooting-stars. But it would be a pity not to mention the so-called Siberian Meteorite of 1908, because the evidence now indicates that it was not an ordinary meteorite at all.

The impact occurred on 30 June 1908, in the Tunguska region of Siberia. A brilliant object appeared without warning, and became as bright as the Sun, after which there was a violent shock, indicating that something massive had fallen. There were various witnesses—one of whom said that the heat was enough to burn the shirt on his back, even though he was forty miles away from the impact-point—but there were no casualties, since the whole region is more or less uninhabited. The local reindeer population suffered, and pine-trees were blown down over an area more than twenty miles in diameter. Had the missile hit a city, the death-roll would have been colossal.

Little information could be obtained at the time, because of the unsettled state of affairs in Russia, and it was not until 1927 that an expedition went to Tunguska to investigate. It was led by Leonid Kulik, of Leningrad, who had taken a special interest in the problem. It seemed likely that the explosion had been caused by a meteorite, and there were, after all, cases of known meteorite craters—the most famous being that in Arizona, not far from the town of Winslow. The Arizona crater is almost a mile across, and many meteorite fragments have been found in the area.

To his surprise, Kulik found no meteorite material at all. There was plenty of evidence of devastation, but nothing in the form of cosmical débris. This was a puzzle, and led to various remarkable suggestions; flying saucer believers later took the matter up, and proposed that the 'meteorite' was really a visiting space-ship which had blown up on landing!

However, the probable answer is that the missile was a small comet—a theory proposed in the early 1930s by F. J. W. Whipple, Director of the Kew Observatory in London, and the Russian astronomer I. Astapovich. The ices making up the head of the comet would be evaporated by the heat generated on impact, and would produce the shock-wave which blew the trees flat, while the cometary gases would simply dissipate in the atmosphere. After a short period all trace of the comet would be gone. Recent studies by C. P. Florensky and other scientists in the Soviet Union confirm this idea. Expeditions still go to the site; the last was dispatched as recently as 1976.

The Siberian explosion remains the only likely cometary impact in recorded times, and it is not probable that another will happen in the foreseeable future. This is fortunate, because although a comet could not destroy the Earth it could certainly wipe out a city. But all in all there is no reason to be apprehensive, and the danger of being harmed by a comet is much less than that of being knocked down by a car every time you cross a country road.

*Chapter Nine*

# WHENCE COME THE COMETS?

A COMET IS A GHOST-LIKE thing. There is nothing ponderous about it—and, like a ghost, it can vanish without trace. Where, then, does a comet originate?

This brings us on to the question of the origin of the Solar System itself, about which there have been many theories. I do not propose to go into detail here, except to say that the old idea, according to which the planets were pulled off the Sun by the action of a passing star, has been investigated and found to be wanting. (Like so many plausible-sounding theories, it is mathematically untenable.) It is overwhelmingly likely that the Sun and the planets are condensations from an original cloud of dust and gas. It may be that the Sun itself represents the central part of the cloud; it may be that the cloud used to surround an already-shining Sun, and that the Earth and the other planets built up from the material over a long period of time. At least we are confident that the Earth's age can be fixed at around 4,700 million years (or 4·7 æons), and the same is true of the Moon, as we know from analyses of the lunar samples brought home by the Apollo astronauts and the unmanned Russian probes.

No comet can last for anything like so long as this, if it makes regular returns to the Sun. Therefore, either the shorter-period comets are of more recent origin, or else they have spent so much of their lives in regions far from the Sun that they have been able to avoid wasting away.

Today there are several main theories of cometary origin, so let us deal with them one by one, saving the best until last.

(1) *The Capture Theory.* Comets come from interstellar space, and merely chance to approach the Sun. They are then 'caught' by the gravitational pull of Jupiter or one of the other giant planets, and forced into an elliptical orbit, so remaining as members of the Solar System.

This idea was popular for a long time, but more recent

studies of the movements of comets have shown that they do *not* come from interstellar space; and the capture theory has been virtually abandoned.

(2) *The Giant Planet Theory*. Toward the end of the eighteenth century, the great French mathematician Joseph Lagrange suggested that comets might have been shot out of the giant planets. In those days Jupiter and Saturn were believed to be hot bodies, perhaps like miniature suns, and there seemed no reason to doubt that the same could be true of Uranus, discovered in 1781 (Neptune, as we have noted, followed in 1846). This idea of miniature suns was not disproved until the 1920s. It is, of course, true that the internal heat of a giant planet is considerable, and may reach several tens of thousands of degrees at least in the case of Jupiter, but the outer gases are very cold. There is a fundamental difference between a star, which is self-luminous, and a giant planet, which is not.

About a hundred years ago the English astronomer R. A. Proctor worked out the ejection theory in detail. He based his ideas on the fact that many of the short-period comets have their aphelia at roughly the distance of Jupiter's orbit, and concluded that the comets originated inside Jupiter itself. He regarded the famous Great Red Spot as a kind of super-volcano, puffing out comets regularly and so maintaining the supply. Short-period comets which disintegrated would be soon replaced by new ones.

Not many people agreed with Proctor, and little more was heard about the theory until 1953, when it was rather unexpectedly revived by S. K. Vsekhsviatskii, of Kiev Observatory. One of the main objections has always been the great force needed to hurl the comet-making material away from Jupiter, whose escape velocity is thirty-seven miles per second—as against a mere seven miles per second for the Earth. When Vsekhsviatskii first put forward his hypothesis, he suggested that the 'Jovian family' of short-period comets had been ejected from Jupiter, and the longer-period comets from Saturn, Uranus and Neptune respectively; later he amended the theory, and proposed that comets could have been ejected from the four large Jovian satellites, Io, Europa, Ganymede and Callisto. Frankly, this sounds unlikely in the highest degree. Even

Ganymede and Callisto, the largest of the four, are no more than comparable with the smallest planet, Mercury; Io is only slightly larger than our Moon, and Europa is smaller. They do not appear to be active worlds, as has been confirmed by the data sent back by Pioneer 10 in 1973 and Pioneer 11 in 1974. Moreover, there are serious mathematical objections to the whole idea of cometary material being ejected from either a planet or a satellite, and modern astronomers in general are decidedly unimpressed. So let us pass on to—

(3) *The Collection Theory.* This is my own term for the hypothesis first proposed by Laplace, contemporary and fellow-countryman of Lagrange. Laplace believed that comets originated in an interstellar cloud which was captured long ago by the Sun. The modern version has been developed by R. A. Lyttleton, who, as we have noted, regards comets as flying gravel-banks.

On the collection theory, comets are produced when the Sun passes through an interstellar cloud, and there is a sort of 'gravitational lens' effect, concentrating dust and frozen gases in the area opposite to the Sun's motion through the cloud. Lyttleton reasons that our familiar comets must be recent acquisitions, and he cites the case of Halley's Comet, which loses a certain amount of mass every seventy-six years when it returns to perihelion. Reckoning backwards, so to speak, it can be calculated that a mere ten million years ago Halley's Comet would have been as massive as the Earth if it had followed its present orbit ever since. This is clearly absurd, so that Halley's Comet cannot have been suffering this steady wastage for anything like so long as ten million years.

There is nothing implausible in the suggestion that the Sun's latest encounter with a dust-and-gas cloud has been relatively recent, but objections to the collection theory come from mathematicians who maintain that comets do not move in quite the way that they would be expected to do if they were produced as Lyttleton believes. One modification (not Lyttleton's) suggests that the collected material forms into a dust-cloud, making a zone round the Sun from which comets are built up. This leads us on to the next hypothesis:

(4) *Oort's Cloud.* Immanuel Kant, the eighteenth-century

German philosopher, had the germ of an idea when he wrote that comets are formed in remote regions, and are made up of particles 'of the lightest material there is'. Suppose, then, that there is a 'cloud' of comets, moving round the Sun at an immense distance—a sort of refrigerated cometary reservoir? This was proposed in 1930 by Ernst Öpik, and worked out in more detail twenty years later by J. H. Oort in Holland.

According to Oort, comets are composed of the material left over after the formation of the main bodies of the Solar System. Normally they travel round the Sun in approximately circular paths, moving very slowly (perhaps only an inch or two per second), and remaining well beyond the range of our telescopes; in fact, what has become known as Oort's Cloud may extend out almost half-way to the nearest star. There may be upward of a hundred million comets in the cloud, and most of them remain in stable orbits. However, perturbations by any passing star—or, perhaps, by mutual collisions between the particles themselves—could cause a change, putting the affected comet into a path which would send it hurtling toward the Sun. At first the movement would be gradual, but as the distance lessened the velocity would increase. When the comet approached the Sun, it would be moving at a tremendous rate; it would swing past perihelion and then return to the outer regions from which it had come—unless it encountered a planet, and was forced into a short-period orbit.

This theory would explain many of the facts. It would show why the short-period comets are faint; they are losing some of their material at each return to perihelion, whereas the comets with very long periods (hundreds, thousands or even millions of years*) come back so seldom that they have not wasted away. It would also explain why bright comets tend to appear at intervals of a relatively few years, after which there is a long interval before any more are seen. The Sun-grazers of the nineteenth century could well have been sent on their way from the same part of Oort's Cloud as a result of the same perturbing influence. It has also been suggested

* The longest period computed for any comet is 24,000,000 years for Delavàn's Comet, 1914V. G. van Biesbroeck, who made the calculation, added that the aphelion distance was around 15,810 thousand million miles from the Sun. No doubt this result is of the right order, even though we cannot expect it to be at all precise.

that there may even be two clouds: one at about twice the distance of Neptune, and the other much more remote, occupying the zone between 30,000 and 100,000 astronomical units from the Sun. (The nearest known star, Proxima Centauri, is roughly 270,000 astronomical units away.) The middle of the main zone would therefore be about one light-year from the Sun.

Yet here too there are objections, and some astronomers have no faith in a comet-cloud of any description. First, how did the comets get there? If they were formed in what we may call the middle part of the Solar System, where Jupiter now moves, they could possibly have been hurled outward by Jupiter's gravitational power and then forced into remote, near-circular orbits by the pulls of nearby stars; but this involves some very special assumptions. Lyttleton has also stressed that there are simply not enough stars sufficiently close to wrench numbers of comets away from the cloud and send them inward. Any normal stars lying within a few light-years of us would undoubtedly have been found long ago, and to suppose that there are quantities of dead, non-luminous stars nearer to us than Proxima Centauri is straining the possibilities.

Obviously, then, our ideas about the origin of comets are still in a state of flux. Cloud or no cloud? Gravel-banks or dirty ice-balls—or something different from either? It would help immensely if it were possible to send an unmanned space-craft to (or, better, through) a comet to see just what is to be found there. In fact such a trip is far from easy, even in this age of travel to the Moon and rockets to the giant planets. One scheme is to contact a short-period comet whose orbit is not sharply tilted, and which comes reasonably close to the Earth; Finlay's Comet and D'Arrest's Comet have been suggested. There are also tentative plans to send a probe out to Encke's Comet. The obvious candidate—Halley's Comet—poses even greater problems, because of its retrograde motion. One proposal involves the gravitational field of Saturn (Fig. 21); if the probe were swung round Saturn in 1983, it could conceivably be hurled back into a retrograde orbit, so that it could 'catch up' its target even before Halley's Comet comes back into telescopic range. There has even been mention of a

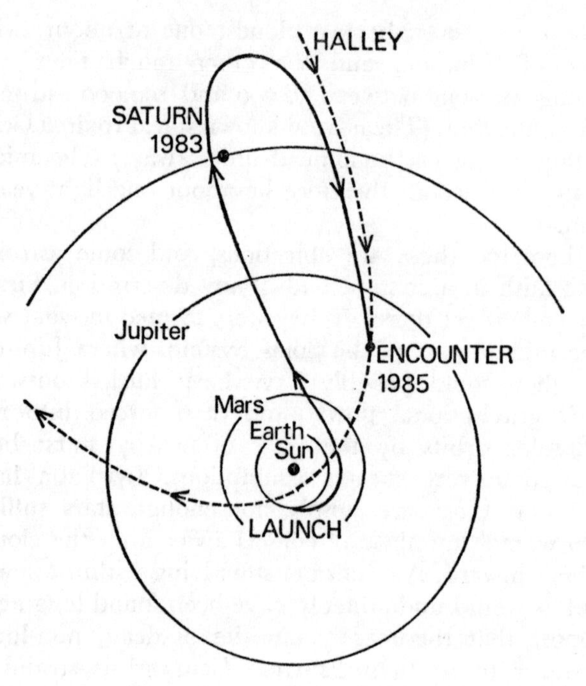


Fig. 21. Hypothetical space probe to Halley's Comet, making use of the gravitational pull of Saturn. The probe would be swung around Saturn in 1983 and would then move back toward the sun in a retrograde direction, catching up to the comet in 1985. It is not very likely that the experiment will be tried at the coming return of the comet, because the launching would have to take place in the near future and no definite plans for this have been made

two-comet probe, visiting both Halley and Giacobini-Zinner; and designs for 'solar sails' have been published. All this is admittedly speculative at the moment—but much less so than reaching the Moon was in, say, 1940.

I am well aware that this survey of cometary astronomy is very much of an open-ended story. I hope, however, that I have managed to convey something of its fascination, and that you will take more than a casual interest next time you see one of these strange, insubstantial visitors from outer space.

# APPENDICES

*Appendix I*

## PERIODICAL COMETS WHICH HAVE BEEN OBSERVED AT MORE THAN ONE RETURN

| *Comet* | *Period years* | *Distance from Sun (Astronomical Units) Perihelion* | *Aphelion* | *Eccentricity* | *Inclination* | *Number of Returns* | *Last Return* |
|---|---|---|---|---|---|---|---|
| Encke | 3·3 | 0·34 | 4·09 | 0·85 | 12·0 | 51 | 1977 |
| Grigg-Skjellerup | 5·1 | 1·00 | 4·94 | 0·66 | 21·1 | 13 | 1977 |
| Tempel 2 | 5·3 | 1·36 | 4·68 | 0·55 | 12·5 | 16 | 1977 |
| Honda-Mrkós-Ṕajdusáková | 5·3 | 0·58 | 5·49 | 0·58 | 13·1 | 5 | 1974 |
| Neujmin 2 | 5·4 | 1·34 | 4·84 | 0·57 | 10·6 | 2 | 1927 |
| Tempel 1 | 5·5 | 1·50 | 4·73 | 0·52 | 10·5 | 6 | 1977 |
| Tuttle-Giacobini-Kresák | 5·6 | 1·15 | 5·13 | 0·63 | 13·6 | 5 | 1973 |
| Tempel-Swift | 5·7 | 1·15 | 5·22 | 0·64 | 5·4 | 4 | 1908 |
| Wirtanen | 5·9 | 1·26 | 5·16 | 0·61 | 12·3 | 5 | 1974 |
| D'Arrest | 6·2 | 1·17 | 5·61 | 0·66 | 16·7 | 13 | 1976 |
| Du Toit-Neujmim-Delporte | 6·3 | 1·68 | 5·15 | 0·51 | 2·9 | 2 | 1970 |
| Di Vico-Swift | 6·3 | 1·62 | 5·21 | 0·52 | 3·6 | 3 | 1965 |
| Pons-Winnecke | 6·3 | 1·25 | 5·61 | 0·64 | 22·3 | 18 | 1976 |
| Forbes | 6·4 | 1·53 | 5·36 | 0·56 | 4·6 | 5 | 1974 |
| Kopff | 6·4 | 1·57 | 5·34 | 0·55 | 4·7 | 11 | 1976 |
| Schwassmann-Wachmann 2 | 6·5 | 2·14 | 4·83 | 0·39 | 3·7 | 8 | 1974 |
| Giacobini-Zinner | 6·5 | 0·99 | 5·98 | 0·71 | 31·7 | 9 | 1972 |

| | | | | | | | |
|---|---|---|---|---|---|---|---|
| Churyumov-Gerasimenko | 6·6 | 1·30 | 3·51 | 0·63 | 7·12 | 2 | 1976 |
| Wolf-Harrington | 6·6 | 1·52 | 5·38 | 0·54 | 18·4 | 6 | 1977 |
| Tsuchinshan 1 | 6·6 | 1·49 | 5·57 | 0·58 | 10·5 | 3 | 1977 |
| Perrine-Mrkóe | 6·7 | 1·27 | 5·85 | 0·64 | 17·8 | 6 | 1974 |
| Reinmuth 2 | 6·7 | 1·94 | 5·19 | 0·46 | 7·0 | 5 | 1974 |
| Borrelly | 6·8 | 1·32 | 5·84 | 0·63 | 30·2 | 9 | 1974 |
| Johnson | 6·8 | 2·20 | 4·96 | 0·39 | 13·9 | 5 | 1976 |
| Tsuchinshan 2 | 6·8 | 1·78 | 5·40 | 0·51 | 6·7 | 2 | 1971 |
| Harrington | 6·8 | 1·58 | 5·60 | 0·56 | 8·7 | 2 | 1960 |
| Gunn | 6·8 | 2·45 | 4·74 | 0·32 | 10·4 | — | constant |
| Arend-Rigaux | 6·8 | 1·44 | 5·76 | 0·60 | 17·8 | 5 | 1977 |
| Brooks 2 | 6·9 | 1·84 | 5·39 | 0·49 | 5·6 | 11 | 1974 |
| Finlay | 6·9 | 1·10 | 6·19 | 0·70 | 3·6 | 9 | 1974 |
| Taylor | 7·0 | 1·95 | 3·64 | 0·47 | 20·6 | 2 | 1977 |
| Holmes | 7·0 | 2·16 | 5·20 | 0·41 | 19·2 | 5 | 1972 |
| Daniel | 7·1 | 1·66 | 5·72 | 0·55 | 20·1 | 5 | 1964 |
| Harrington-Abell | 7·2 | 1·77 | 5·68 | 0·52 | 16·8 | 4 | 1976 |
| Shajn-Schaldach | 7·3 | 2·23 | 5·28 | 0·41 | 6·2 | 2 | 1971 |
| Faye | 7·4 | 1·62 | 5·98 | 0·58 | 9·1 | 17 | 1976 |
| Ashbrook-Jackson | 7·4 | 2·29 | 5·33 | 0·40 | 12·5 | 4 | 1971 |
| Whipple | 7·5 | 2·48 | 5·16 | 0·35 | 10·2 | 6 | 1970 |
| Reinmuth 1 | 7·6 | 2·00 | 5·76 | 0·49 | 8·3 | 6 | 1973 |
| Arend | 7·9 | 1·84 | 4·00 | 0·54 | 20·0 | 4 | 1975 |
| Oterma | 7·9 | 3·39 | 4·53 | 0·14 | 4·0 | 3 | 1958 |
| Schaumasse | 8·2 | 1·20 | 6·92 | 0·70 | 12·0 | 7 | 1976 |

| *Comet* | *Period years* | *Distance from Sun (Astronomical Units)* | | *Eccentricity* | *Inclination* | *Number of Returns* | *Last Return* |
|---|---|---|---|---|---|---|---|
| | | *Perihelion* | *Aphelion* | | | | |
| Jackson-Neujmin | 8·4 | 1·43 | 6·83 | 0·65 | 14·1 | 2 | 1970 |
| Wolf | 8·4 | 2·52 | 5·78 | 0·40 | 27·3 | 12 | 1975 |
| Smirnova-Chernykh | 8·5 | 3·60 | 4·20 | 0·15 | 6·6 | — | constant |
| Comas Solá | 8·6 | 1·77 | 6·59 | 0·58 | 13·4 | 6 | 1969 |
| Kwerns-Kwee | 9·0 | 2·23 | 6·43 | 0·49 | 9·0 | 2 | 1972 |
| Swift-Gehrels | 9·3 | 1·36 | 4·40 | 0·69 | 9·3 | 2 | 1972 |
| Neujmin 3 | 10·6 | 1·98 | 7·66 | 0·59 | 3·9 | 3 | 1972 |
| Gale | 11·0 | 1·18 | 8·70 | 0·76 | 11·7 | 2 | 1938 |
| Klemola | 11·0 | 1·76 | 4·94 | 0·64 | 10·6 | 2 | 1976 |
| Väisälä 1 | 11·3 | 1·87 | 8·19 | 0·63 | 11·5 | 4 | 1971 |
| Slaughter-Burnham | 11·6 | 2·54 | 7·72 | 0·50 | 8·2 | 2 | 1970 |
| Van Biesbroeck | 12·4 | 2·40 | 5·35 | 0·55 | 6·62 | 2 | 1966 |
| Wild | 13·3 | 1·98 | 9·24 | 0·65 | 19·9 | 2 | 1973 |
| Tuttle | 13·8 | 1·02 | 10·46 | 0·82 | 54·4 | 9 | 1967 |
| Du Toit 1 | 15·0 | 1·29 | 10·85 | 0·79 | 18·7 | 2 | 1974 |
| Schwassmann-Wachmann 1 | 15·0 | 5·45 | 6·73 | 0·11 | 9·7 | — | constant |
| Neujmin 1 | 17·9 | 1·54 | 12·16 | 0·78 | 15·0 | 4 | 1966 |
| Crommelin | 27·9 | 0·74 | 17·65 | 0·92 | 28·9 | 4 | 1956 |
| Tempel-Tuttle | 32·9 | 0·98 | 19·56 | 0·90 | 162·7 | 4 | 1965 |
| Stephan-Oterma | 38·8 | 1·60 | 21·34 | 0·86 | 17·9 | 2 | 1942 |
| Olbers | 69·5 | 1·18 | 32·62 | 0·93 | 44·6 | 3 | 1956 |

| | | | | | | | |
|---|---|---|---|---|---|---|---|
| Pons-Brooks | 71·0 | 0·77 | 33·51 | 0·96 | 74·2 | 3 | 1954 |
| Brorsen-Metcalf | 71·9 | 0·49 | 34·11 | 0·97 | 19·2 | 2 | 1919 |
| Halley | 76·1 | 0·59 | 35·33 | 0·97 | 162·2 | 27 | 1910 |
| Herschel-Rigollet | 156·0 | 0·75 | 56·94 | 0·97 | 64·2 | 2 | 1939 |
| Grigg-Mellish | 164·3 | 0·92 | 58·00 | 0·97 | 109·8 | 2 | 1907 |

It must be remembered that these elements alter considerably for each revolution. The values given here apply to the beginning of 1977. Du Toit 1 (Comet 1944 III) was reported again in 1974, but the observations cannot be regarded as definite.

Orbits with inclinations over 90 degrees indicate retrograde motion. There are three comets in the list with retrograde orbits: Tempel-Tuttle, Halley and Grigg-Mellish.

## LOST PERIODICAL COMETS

| | | | | | | | |
|---|---|---|---|---|---|---|---|
| Brorsen | 5·5 | 0·59 | 5·61 | 0·81 | 29·4 | 5 | 1879 |
| Biela | 6·6 | 0·86 | 6·19 | 0·76 | 12·6 | 6 | 1852 |
| Westphal | 61·9 | 1·25 | 30·03 | 0·92 | 40·9 | 2 | 1913 |

It seems certain that these comets have disintegrated.

## PERIODICAL COMETS SEEN AT ONLY ONE RETURN

There are various reasons for 'losing' comets. A few, listed separately, seem definitely to have disintegrated. In other cases the orbit has been so violently perturbed that we have lost contact; the classic case is that of Lexell's Comet of 1770. The list also includes several short-period comets which have been discovered recently, and which will no doubt return on schedule; these are identified by an asterisk.

| *Comet* | *Period in years* | *Last seen* |
|---|---|---|
| Wilson-Harrington | 2·3 | 1949 |
| Helfenzrieder | 4·5 | 1766 |
| Blanpain | 5·1 | 1819 |
| Du Toit 2 | 5·3 | 1945 |
| La Hire | 5·4 | 1678 |
| Barnard 1 | 5·4 | 1884 |
| Schwassmann-Wachmann 3 | 5·4 | 1930 |
| Grischow | 5·4 | 1743 |
| Clark* | 5·6 | 1973 |
| Brooks 1 | 5·6 | 1886 |
| Lexell | 5·6 | 1770 |
| Kulin | 5·6 | 1939 |
| Pigott | 5·9 | 1783 |
| West-Kohoutek-Ikemura* | 6·1 | 1975 |
| Kohoutek* | 6·2 | 1975 |
| Spitaler | 6·4 | 1890 |
| Harrington-Wilson | 6·4 | 1951 |
| Barnard 3 | 6·6 | 1892 |
| Giacobini | 6·7 | 1896 |
| Schorr | 6·7 | 1918 |
| Longmore* | 7·1 | 1974 |
| Swift 2 | 7·2 | 1895 |
| Denning 2 | 7·4 | 1894 |
| Metcalf | 7·9 | 1906 |
| Gehrels 2* | 7·9 | 1973 |
| Gehrels 3* | 8·4 | 1975 |
| Denning 1 | 8·7 | 1881 |
| Swift 1 | 8·9 | 1889 |
| Boethin* | 11·0 | 1975 |
| Peters | 13·4 | 1846 |
| Gehrels 1* | 14·6 | 1973 |
| Perrine | 16·4 | 1916 |
| Pons-Gambart | 63·9 | 1827 |
| Ross | 64·6 | 1883 |
| Dubiago | 67·0 | 1921 |
| Di Vico | 75·7 | 1846 |
| Väisälä 2 | 85·5 | 1942 |
| Swift-Tuttle | 119·6 | 1862 |
| Barnard 2 | 128·3 | 1889 |
| Mellish | 145·3 | 1917 |

*Appendix II*

# SOME NOTABLE COMETS SEEN SINCE 1680

1680 Brilliant comet, visible in daylight. Discovered by G. Kirch.

1744 De Chéseaux' six-tailed comet (actually discovered by Klinkenberg).

1811 Magnificent comet discovered by H. Flaugergues. Largest coma ever recorded (1¼ million miles in diameter).

1843 Brilliant daylight comet, superior to that of 1811. Longest recorded tail (200,000,000 miles).

1858 Donati's Comet; beautiful curved main tail.

1861 Brilliant comet, discovered by Tebbutt. The Earth passed through its tail.

1862 Bright comet, discovered by Lewis Swift.

1874 Coggia's Comet; bright naked-eye object.

1882 Brilliant comet; the first to be well photographed (by Gill).

1901 Bright southern comet, discovered by Paysandu. Yellowish in colour.

1910 Daylight Comet.

1927 Skjellerup's Comet. Bright for a brief period.

1947 Bright southern comet; conspicuous only for a short while.

1948 Bright comet; discovered during a total eclipse of the Sun.

1957 Arend-Roland; not particularly brilliant, but with the interesting 'spike' phenomenon. Moderately conspicuous with the naked eye during evenings in April.

1957 Mrkós' Comet; fairly bright naked-eye object during mornings in autumn. Comparable with Arend-Roland.

1962 Seki-Lines; bright.

1965 Ikeya-Seki; brilliant as seen from parts of the southern hemisphere.

1969 Tago-Sato-Kosaka; bright. Found to be surrounded by a hydrogen cloud.

1970 Bennett; bright naked-eye object with an impressive tail.

1973 Kohoutek's Comet; as a spectacle, promised much more than it achieved.

1976 West's Comet; bright naked-eye object during early mornings in March.

*Appendix III*

# GLOSSARY

THE FOLLOWING GLOSSARY is not intended to be at all complete, and is restricted to terms which have been used frequently throughout this book.

APHELION. The furthest point of a planet or a comet from the Sun during its orbit.
COMA. The most prominent part of a comet—the 'fuzz' surrounding the nucleus.
ELLIPSE. A closed curve which may be described as an 'oval' (though this will not appeal to mathematicians!). The ECCENTRICITY of the ellipse depends upon how 'long and thin' it is. If the eccentricity is zero, the ellipse has become a CIRCLE.
HEAD. That part of a comet which includes both the nucleus and the coma.
HYPERBOLA. A very open curve—more so than a parabola.
ICES. Frozen materials, not necessarily frozen water; for instance, it is possible to have 'ammonia ice'. There are many kinds of 'ices' contained in comets.
METEOR. A small particle travelling round the Sun, becoming visible only when it enters the Earth's atmosphere and is destroyed by friction against the air-particles.
METEORITE. A larger body, which can reach the ground without being destroyed. A meteorite is not simply a large meteor, but is more nearly related to a minor planet.
METEOROID. A term embracing all the meteoritic bodies in space.
MINOR PLANET (or ASTEROID). A very small planet. Most of the asteroids keep to the region between the orbits of Mars and Jupiter, but some have orbits which bring them well away from the main swarm and may sometimes involve close approaches to the Earth.
NUCLEUS. The most massive part of a comet, situated inside the coma.
ORBIT. The path through space of any astronomical body.
PARABOLA. An open curve. If a comet moves in a parabolic orbit, it will never return to the Sun.
PERIHELION. The nearest point of a planet or comet to the Sun during its orbit.
PERIODICAL COMET. A comet which returns regularly to perihelion.

Conventionally, the term is restricted to comets with periods short enough for their returns to be predicted.

PERTURBATION. The disturbance in motion produced by the gravitational pull of one astronomical body upon another.

RETURN. The time when a comet comes back to perihelion.

TAIL. The part of a comet streaming out from the coma; it is made up of very small particles and very tenuous gas, and always points more or less away from the Sun. Not all comets have tails.

# INDEX